ENGAGE IN THE
MATHEMATICAL
PRACTICES

STRATEGIES TO BUILD NUMERACY AND
LITERACY WITH K–5 LEARNERS

KIT NORRIS　　SARAH SCHUHL

Solution Tree | Press

a division of
Solution Tree

555 North Morton Street
Bloomington, IN 47404
800.733.6786 (toll free) / 812.336.7700
FAX: 812.336.7790

email: info@solution-tree.com
solution-tree.com

Visit **go.solution-tree.com/MathematicsatWork** to download the free reproducibles in this book.

Printed in the United States of America

Library of Congress Cataloging-in-Publication Data

Names: Norris, Kit. | Schuhl, Sarah.

Title: Engage in the mathematical practices : strategies to build numeracy
   and literacy with K-5 learners / Kit Norris and Sarah Schuhl.

Description: Bloomington, IN : Solution Tree Press, [2016] | Includes
   bibliographical references and index.

Identifiers: LCCN 2015047571 | ISBN 9781936764761 (perfect bound)

Subjects: LCSH: Numeracy. | Number concept. | Mathematics--Study and teaching
   (Early childhood)

Classification: LCC QA141.15 .N67 2016 | DDC 372.7--dc23 LC record available at http://lccn.loc.gov/2015047571

Solution Tree
Jeffrey C. Jones, CEO
Edmund M. Ackerman, President

Solution Tree Press
President: Douglas M. Rife
Senior Acquisitions Editor: Amy Rubenstein
Editorial Director: Lesley Bolton
Managing Production Editor: Caroline Weiss
Production Editor: Tara Perkins
Copy Editor: Sarah Payne-Mills
Proofreader: Elisabeth Abrams
Text and Cover Designer: Abigail Bowen

We are both forever grateful to our families for their love and understanding of us while we spent time writing this book.

To Jon, Jacob, and Sam, you made this possible and words cannot begin to express my heartfelt appreciation for all you are. Thank you for your daily inspiration and support. You mean the world to me.

—Sarah Schuhl

To Kyle, Alison, Carrie, and Thacher, your support and interest in my work make such a difference to me; and to my grandchildren, Carolyn, Judson, Miller, Grafton, Dean, and Kingsley, thanks for always being willing to play my math games; and to Chip, your guidance, feedback, and love balance and enrich my life.

—Kit Norris

Visit **go.solution-tree.com/MathematicsatWork**
to download the reproducibles in this book.

ACKNOWLEDGMENTS

There are many educators and students who have shaped this work and continue to grow our own understanding of effective mathematics instruction. In addition to our colleagues with the National Council of Supervisors of Mathematics and the National Council of Teachers of Mathematics and the numerous teachers we have worked with across the United States in workshops and in classrooms, we would like to specifically recognize author, mathematics educator, and friend Timothy Kanold. Through his vision and leadership, we have had the opportunity to continually challenge ourselves to improve student learning of mathematics as we work with K–12 mathematics collaborative teams.

We also greatly appreciate our partnership with Solution Tree. With its support, we have been able to work with mathematics teacher teams in large and small settings and create this book. We value their dedication to educators and students as evidenced by their continual encouragement for and support of the work educators are doing in the field.

Several U.S. schools have contributed to this resource. We are grateful for the many opportunities we have had to facilitate rich collaborative conversations through lesson study centered on student engagement, which resulted in a deeper understanding of the strategies needed to emphasize the Mathematical Practices while teaching content. Specifically, we greatly appreciate the willingness from the teachers at the following schools to include our tasks in their lessons and share their students' efforts and insights. These schools include:

- Bentley Elementary School
  Salem, Massachusetts

- E. Pole Elementary School
  Taunton, Massachusetts

- Green Meadow School
  Maynard, Massachusetts

- Fowler School
  Maynard, Massachusetts

- French Valley Elementary School
  Winchester, California

We are appreciative of our many experiences to grow and share practices related to mathematics collaborative teams and instructional strategies to productively engage K–5 learners. We are thankful to you, our readers, for joining us in this journey. Together, we can make a difference in student learning.

Solution Tree Press would like to thank the following reviewers:

Bill Barnes
Curriculum Director
Howard County Public Schools
Ellicott City, Maryland

Judy Curran Buck
Mathematics Education Consultant
Derry, New Hampshire

Amy McQueen
Mathematics Coach
David Douglas School District
Portland, Oregon

Dalila L. Mendoza
Literacy Coach
Elizabeth Pole School
Taunton, Massachusetts

Sue Pippen
Pippen Consulting
Past President, Illinois Council of Teachers of
  Mathematics
Plainfield, Illinois

David Pugalee
Director, Center for Science, Technology,
  Engineering, and Mathematics Education
University of North Carolina at Charlotte
Charlotte, North Carolina

TABLE OF CONTENTS

CHAPTER 3

Standard for Mathematical Practice 3:
Construct Viable Arguments and Critique the Reasoning of Others . . . 63

CHAPTER 4

Standard for Mathematical Practice 4: Model With Mathematics 93

CHAPTER 8
Standard for Mathematical Practice 8:
Look for and Express Regularity in Repeated Reasoning 203

ABOUT THE AUTHORS

Kit Norris is a mathematics consultant focusing on teachers' professional growth and development. She is a frequent speaker at local and national conferences. Kit works with the Bay State Reading Institute helping elementary school teachers understand the parallels between literacy and mathematics best practices. Kit began her career as a middle school mathematics teacher at the Eaglebrook School in Deerfield, Massachusetts. She later became the mathematics department chair and director of the academic program at the Fenn School in Concord, Massachusetts.

As a member of the Mathematics at Work™ team, she supports teachers in implementing high-leverage team actions necessary to improve student learning. Kit is also a member of the National Council of Supervisors of Mathematics (NCSM) and has served on the Board of Directors as the editor for the *Position Papers*. She coauthored *Teaching Today's Mathematics in the Middle Grades*, *Great Tasks for Mathematics K–5*, and *Great Tasks for Mathematics 6–12* and was a contributing author to *It's TIME: Themes and Imperatives for Mathematics Education*.

She is a recipient of the 1994 Presidential Award for Excellence in Teaching Mathematics and the 1998 Klingenstein Fellowship. This fellowship enabled her to pursue research, focusing on strategies for supporting and promoting professional learning opportunities for educators. In 2015, she was inducted into the Mathematics Educators Hall of Fame in Massachusetts.

She has a bachelor's degree from Skidmore College and a master's degree in educational leadership from Columbia University. Kit and her husband live outside Boston and enjoy keeping up with their grandchildren and playing the occasional round of golf.

Sarah Schuhl began her education career in 1994 and spent nearly twenty years working first as a secondary mathematics teacher in Oregon and California, then as a high school instructional coach, and then serving as a K–12 mathematics specialist in the Centennial School District in Oregon. As an instructional coach, she was instrumental in the creation of professional learning communities, working with teachers to make large gains in student achievement. She earned the district's Triple C Award in 2013.

Sarah now works with teachers in schools and districts as a K–12 mathematics and multisubject educational consultant. Through short- and long-term professional development and coaching, she works with collaborative teams and individual teachers to understand and implement content and process standards, create and use common assessments and high-level tasks, design lessons, and analyze and respond to student learning as evidenced in data. Her work with the Oregon Department of Education includes designing mathematics assessment items, test specifications and blueprints, and rubrics for achievement-level descriptors. She also works with schools in need of improvement to increase student learning.

From 2010 to 2013, Sarah served as a member and chair of the National Council of Teachers of Mathematics Editorial Panel for the journal *Mathematics Teacher*. She is also a contributing author to *Core Focus on Math*, a Common Core middle school mathematics textbook series, as well as coauthor on Common Core intervention mathematics materials for grades 3–5, *Digging Into Math*.

Sarah earned a bachelor of science in mathematics from Eastern Oregon University and a master of science in teaching mathematics from Portland State University.

To book Kit Norris or Sarah Schuhl for professional development, contact pd@solution-tree.com.

INTRODUCTION

The art of teaching is the art of assisting discovery.

—MARK VAN DOREN

Elementary teachers—we salute you!

Teaching is a complex craft. As an elementary teacher, expectations of you continue to increase as resources continue to dwindle. Not only do you need to understand the content and practices you are teaching in mathematics, but you are also expected to deeply know other disciplines. You must be cognizant of early childhood development theories and mindful of all stakeholders with whom you must communicate—administration, colleagues, parents, and students. Your days with students are varied, action packed, unpredictable, rewarding, challenging, and perhaps, exhausting.

Our book is written with you in mind. Our purpose is to support your work as a mathematics teacher as you seek to provide mathematics students with experiences that stimulate them to think, analyze, conjecture, reason, model, and make connections between and among topics. We also want to help you deepen the connections between literacy and mathematics. Best practices in one discipline transfer to best practices in another.

Student Engagement and the Standards for Mathematical Practice

It is time to reimage the structure of the mathematics block. Whole-group instruction directed from the front of the room *for an entire lesson* needs to disappear. Learners must be engaged in the process of *doing* mathematics in order for them to learn and retain their understandings. Sitting and watching a teacher explain and demonstrate procedures and steps is like learning how to play a sport by watching a video. As described in *Principles to Actions*, mathematics learning is an active process "in which each student builds his or her own mathematical knowledge from personal experiences, coupled with feedback from peers, teachers and other adults, and themselves" (National Council of Teachers of Mathematics [NCTM], 2014, p. 9). NCTM continues by identifying specific experiences for learners that enable them to:

- engage with challenging tasks that involve active meaning making and support meaningful learning;

- connect new learning with prior knowledge and informal reasoning and, in the process, address preconceptions and misconceptions;

- acquire conceptual knowledge as well as procedural knowledge, so that they can meaningfully organize their knowledge, acquire new knowledge, and transfer and apply knowledge to new situations;

- construct knowledge socially, through discourse, activity, and interaction related to meaningful problems;

- receive descriptive and timely feedback so that they can reflect on and revise their work, thinking, and understanding; and

- develop metacognitive awareness of themselves as learners, thinkers, and problem solvers, and learn to monitor their learning and performance. (NCTM, 2014, p. 9)

Note that every bulleted item remains focused on the learner's active engagement. How do you as teachers provide specific experiences so that learners take those actions? Consider the emerging importance of students solving problems.

What once was considered problem solving is now understood as a simple exercise in which students practice skills. While this is important, students must move beyond the practice of procedures and skills to engagement in answering meaningful questions. These questions, often referred to as *high-cognitive-demand tasks*, present students with a situation in which the solution itself or the path to the solution is not easily determined. Making sense of the problem, trying varying approaches, reflecting on the selected approach along the way, and evaluating whether the answer makes sense are part of the process as students work with high-cognitive-demand tasks. As Glenda Lappan and Diane Briars (1995) point out:

> There is no decision that teachers make that has a greater impact on students' opportunities to learn and on their perceptions about what mathematics is than the selection or creation of the tasks with which the teacher engages students in studying mathematics. (p. 139)

So what are the mathematical tasks you choose to use each day with students? How do the tasks show connections in student learning? How do students access and build the stamina needed to solve the tasks?

To help answer these questions, we describe engaging strategies throughout this book that you can use in your classroom to develop the critical reasoning skills necessary for all students to learn mathematical content. We've centered these strategies on the eight Common Core Standards for Mathematical Practice (National Governors Association Center for Best Practices [NGA] & Council of Chief State School Officers [CCSSO], 2010), which illustrate the habits of mind students must develop to successfully solve problems.

1. Make sense of problems and persevere in solving them.

2. Reason abstractly and quantitatively.

3. Construct viable arguments and critique the reasoning of others.

4. Model with mathematics.

5. Use appropriate tools strategically.

6. Attend to precision.

7. Look for and make use of structure.

8. Look for and express regularity in repeated reasoning. (NGA & CCSSO, 2010, pp. 6–8)

Figure I.1 illustrates the relationships among the Standards for Mathematical Practice.

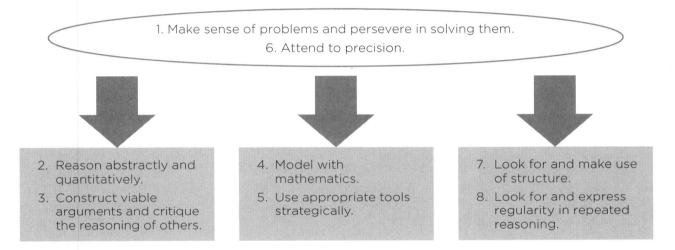

Source: Adapted from McCallum, 2011.

Figure I.1: Relationships among the Standards for Mathematical Practice.

Students begin by making sense of meaningful tasks and pursuing a solution in spite of possible setbacks along the way (Mathematical Practice 1). You'll note the sixth practice, attend to precision, is also in the overarching oval: as mathematically proficient students engage in finding solutions that make sense, they also focus on the precision of their work, the appropriate use of units and labels, and the clarity of their communication. As students grapple with a meaningful task, they have opportunities to engage in the other Mathematical Practices. For instance, Mathematical Practices 2 and 3 focus on students reasoning and communicating with one another, while Mathematical Practices 4 and 5 can be categorized as utility standards, those habits of mind requiring the use of mathematical models and tools to make sense of the world and apply learning. Mathematical Practices 7 and 8 feature the student use of mathematical patterns through structure and repeated reasoning. Together, these eight habits of mind define the practices students must develop to be successful mathematics learners.

These practices are not new; they did not first appear in the Common Core State Standards (CCSS). Rather, they have been an expectation for mathematics learning since the 1980s. The Common Core State Standards simply once again verified their importance to student learning. These practices are steeped in research and represent a blend of two pivotal and foundational resources: (1) *Principles and Standards for School Mathematics* (NCTM, 2000) and (2) the National Research Council's *Adding It Up: Helping Children Learn Mathematics* (Kilpatrick, Swafford, & Findell, 2001). While we acknowledge some states have not adopted the Common Core State Standards and are, therefore, not calling their mathematical process standards the *Mathematical Practices*, each state has process standards derived from these sources. This book will reference the Mathematical Practices, knowing these are parallel in structure to any mathematical process standards as the habits of mind students need to develop to be proficient mathematicians. For additional background information on the Common Core State Standards, see appendix C (page 251).

Teacher Collaboration and Instruction

Teachers are stronger and more effective when they can build shared knowledge with their colleagues. In *Principles to Actions*, NCTM (2014) recognizes the importance of collaborative teams in its professionalism guiding principle: "Mathematics teachers are professionals who do not do this work in isolation. They cultivate and support a culture of professional collaboration and continual improvement, driven by an abiding sense of interdependence and collective responsibility" (p. 99).

Collaborative teams work together to answer the four critical professional learning community questions in *Learning by Doing* (DuFour, DuFour, Eaker, & Many, 2010).

1. What do we want all students to know and be able to do?

2. How will we know if they know it?

3. How will we respond if they don't know it?

4. How will we respond if they do know it?

Answering these questions thoughtfully as a collaborative team focused on data and results requires reflection, a willingness to learn and share ideas, and a relentless focus on student results. Together, you erase inequities in student learning from teacher to teacher and create common expectations of student learning as a collective focus to achieve.

In *Visible Learning for Teachers*, John Hattie (2012) states:

> Planning can be done in many ways, but the most powerful is when teachers work together to develop plans, develop common understandings of what is worth teaching, collaborate on understanding their beliefs of challenge and progress, and work together to evaluate the impact of their planning on student outcomes. (p. 37)

The strategies in this book will be more effective when you and your colleagues determine why to use each and what to look for collectively to ensure students are learning the Standards for Mathematical Practice as well as the content standards. Be sure your team considers connections between mathematics and literacy to enhance instruction and has a shared definition for high-level tasks as you plan lessons collaboratively. And once lessons are planned, it's important to participate in walkthroughs to ensure quality and enhance brainstorming for future lessons.

Connections to Literacy

Your work as elementary educators is complex. How can you blend and incorporate the literacy strategies with content experienced in mathematics? Consider specific strategies your team uses when teaching students to read and write; for instance, small-group instruction, reciprocal teaching, anchor charts, sentence frames, think-alouds, and requests for evidence are just a few actions you likely use during literacy instruction. Fortunately, these best practices in literacy are also effective in teaching and learning mathematics.

Take a moment to use the tool in figure I.2 to jot down some of your favorite instructional strategies to teach literacy. Then, reflect on your mathematics instruction. How can your team implement the approaches used to engage students in reading and writing during mathematics instruction?

Literacy Strategies	Mathematics Strategies

Figure I.2: Instructional strategies for literacy and mathematics.

*Visit **go.solution-tree.com/MathematicsatWork** for a free reproducible version of this figure.*

Throughout this resource, you will see explicit connections between the teaching of literacy and the teaching of mathematics. These connections are noted with a Literacy Connection icon (featured here for your reference) throughout the text.

The Venn diagram in figure I.3 (page 6) compares and contrasts literacy strategies and mathematics strategies. As you read this book, consider using the diagram to record your thinking about the relationships between these two areas. Which strategies do teachers solely use in literacy? Are there strategies that are more appropriate for mathematics? Which strategies can teachers apply to both disciplines?

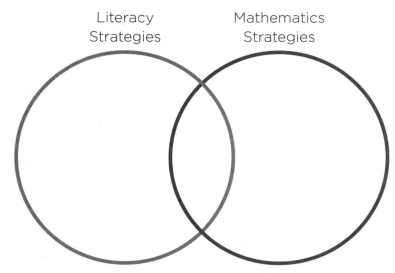

Literacy
Strategies

Mathematics
Strategies

Figure I.3: Literacy and mathematics strategies Venn diagram.

*Visit **go.solution-tree.com/MathematicsatWork** for a free reproducible version of this figure.*

High-Level Tasks

In order to meet the expectations from *Principles to Actions*, students must be presented with questions that engage them in critical thinking. As NCTM (2014) states, "Effective teaching of mathematics engages students in solving and discussing tasks that promote mathematical reasoning and problem solving and allow multiple entry points and varied solution strategies" (p. 10). These types of tasks provide the high-cognitive-demand experiences that enable students to strengthen their ability to meet the expectations of the Mathematical Practices. In addition to promoting reasoning and opportunities for multiple approaches, high-cognitive-demand tasks allow students to deeply understand specific content and make connections within key topics. Such tasks include activities, whole- and small-group work, and a great variety of question formats. As stated in *Beyond the Common Core: A Handbook for Mathematics in a PLC at Work, Grades K–5* (Kanold, 2015a), "In short, the tasks are the problems you choose to determine the pathway of student learning and to assess student success along that pathway" (p. 20).

Within each chapter, we provide tasks that feature the Mathematical Practice under discussion. As you consider implementing these tasks, it is helpful to analyze the level of demand on students. Finding tasks can be time consuming, so we have provided some helpful online resources in appendix G (page 263).

There are several cognitive complexity frameworks educators use to determine the level of reasoning students need to solve a task. Two common frameworks used in all subject areas are Webb's Depth of Knowledge (DOK) model and Bloom's taxonomy. We, however, have chosen to use Margaret Smith and Mary Kay Stein's (2011) framework specific to the complexity of mathematics tasks, which is popular in the mathematics community. Grouping tasks by *lower level* and *higher level* is an easy first step to consider. Smith and Stein's framework focuses on four levels, two groups at the lower level of cognitive demand and two groups at the higher level of demand. Figure I.4 shows a summary of the levels, and appendix F (page 261) offers a more detailed description.

> **Lower-Level Cognitive Demand**
>
> *Memorization:* Requires eliciting information such as a fact, definition, term, or a simple procedure, as well as performing a simple algorithm or applying a formula.
>
> *Procedures without connections:* Requires the engagement of some mental processing beyond a recall of information.
>
> **Higher-Level Cognitive Demand**
>
> *Procedures with connections:* Requires complex reasoning, planning, using evidence, and explanations of thinking.
>
> *Doing mathematics:* Requires complex reasoning, planning, developing, and thinking most likely over an extended period of time.

Source: Smith & Stein, 2012, as cited in Kanold, 2015b, p. 26.

Figure I.4: Four categories of cognitive demand.

To further illustrate lower- and higher-level tasks, consider the following questions in table I.1, representing each level of this framework using grade 2 content.

Table I.1: Examples of Grade 2 Lower-Level- and Higher-Level-Cognitive-Demand Tasks

Lower-Level Cognitive Demand	Higher-Level Cognitive Demand
Memorization: 2 + 5 = ____	*Procedures with connections:* Maria purchased some pears at a farm stand. She also picked 12 apples from the tree in her backyard. She counted 23 pieces of fruit when she got home. How many pears did she purchase?
Procedures without connections: ____ + 4 = 17	*Doing mathematics:* Roberto said that he could represent 38 using base-ten blocks in four ways. Sheila thought that he could only represent 38 with the blocks in 3 different ways. Who is correct? Show how you know.

It is important to remember that students need to experience both high- and low-demand tasks. Within every unit there is certainly a place for practice as students rehearse skills. That practice, however, must be balanced with tasks that require greater opportunities for students to reason, analyze, wonder, critique, compare, and reflect. Such tasks enable students to engage and develop the habits of mind described in the eight Standards for Mathematical Practice.

Lesson Planning

As Kanold (2015c) asserts, "Observations of students learning mathematics should reveal about 60–65 percent of the class time as well-managed peer-to-peer student discourse" (p. 79). Supporting students toward mathematical proficiency takes careful, intentional planning. Such detailed planning is enhanced when teachers have opportunities to work collaboratively.

To highlight the connections among the Mathematical Practices, the selected strategies, and high-level tasks, each chapter features a lesson plan. It asks team members to focus on what they are doing during the lesson *and* what students are doing. Each sample lesson includes the following features.

- **Content standard or learning objective:** The learning objective states the content students will learn in the lesson. You may consider stating learning objectives in terms of "I can" statements. Such statements provide students with direction and a means of reflection at the end of the lesson.

- **Process standard or Standard for Mathematical Practice:** Lessons often include several Mathematical Practices students are simultaneously learning with the content standard. However, each lesson shared at the end of a chapter focuses on one Mathematical Practice and describes the specific actions students will take to learn and demonstrate that practice.

- **Formative assessment process:** The formative assessment process identifies those specific ways students will demonstrate their understanding with opportunities to act on feedback from you or peers during the lesson.

- **Assessing and advancing questions:** This part of the lesson plan offers suggestions for differentiating tasks throughout the lesson. The assessing questions help students who may have difficulty beginning a specific task. These questions help students clarify expectations, review known facts, and clarify ideas. The advancing questions push students' thinking further. These questions offer extensions, link to mathematical ideas, and provide opportunities for further exploration.

- **Beginning-of-class routines:** Each lesson begins with a warm-up or introduction that requires students to investigate or make connections to the lesson's learning objective.

- **Two or three tasks addressing the learning objective:** Tasks usually begin by asking students to consider the question independently and then work with a partner or within a team consisting of four students. Students are challenged to track their thinking and reflect on whether their procedures and possible solutions make sense. These tasks engage students in the Standards for Mathematical Practice while learning content.

- **Closure:** These questions allow *students* to summarize their learning, anticipate a possible next step, or demonstrate their understanding by solving a specific problem independently.

The lesson-planning tool used to create these lessons is a template (figure I.5) from the *Beyond the Common Core: A Handbook for Mathematics in a PLC at Work* series (Kanold, 2012c, 2015a, 2015b, 2015c). (Visit **go.solution-tree.com/MathematicsatWork** to download a free reproducible version of the tool.)

The time some of the lessons require may extend beyond one mathematics block. Feel free to select tasks or adapt the lessons in order to meet the needs of the learners in your classroom. But use caution to make sure *all* students are learning through high-level tasks.

Unit:	Date:	Lesson:

Learning target: As a result of today's class, students will be able to _____.

Formative assessment: How will students be expected to demonstrate mastery of the learning target during in-class checks for understanding?

Probing Questions for Differentiation on Mathematical Tasks	
Assessing Questions (Create questions to scaffold instruction for students who are "stuck" during the lesson or the lesson tasks.)	**Advancing Questions** (Create questions to further learning for students who are ready to advance beyond the learning target.)

Targeted Standard for Mathematical Practice:

Which Mathematical Practice will be targeted for proficiency development during this lesson?

Tasks (Tasks can vary from lesson to lesson.)	What Will the Teacher Be Doing? (How will the teacher present and then monitor student response to the task?)	What Will the Students Be Doing? (How will students be actively engaged in each part of the lesson?)
Beginning-of-Class Routines How does the warm-up activity connect to students' prior knowledge, or how is it based on analysis of homework?		
Task 1 How will the students be engaged in understanding the learning target?		
Task 2 How will the task develop student sense making and reasoning?		
Task 3 How will the task require student conjectures and communication?		
Closure How will student questions and reflections be elicited in the summary of the lesson? How will students' understanding of the learning target be determined?		

Source: Kanold, 2012c. Used with permission.

Figure I.5: CCSS Mathematical Practices lesson-planning tool.

*Visit **go.solution-tree.com/MathematicsatWork** for a free reproducible version of this figure.*

Lesson Study and Walkthroughs

After all of the lesson planning, how will you know if students are engaged, thinking, and working? How will you determine whether they are learning the Standards for Mathematical Practice as well as the content standards? In addition to your own anecdotal evidence and intentional common assessment items, consider using lesson studies and walkthroughs as a collaborative team to build shared knowledge related to what engagement in a practice looks and sounds like during a lesson, especially since the practices are so intertwined.

In describing the meaning of visible teaching and learning, John Hattie (2012) states:

> It is teachers seeing learning through the eyes of students, and students seeing teaching as the key to their ongoing learning. The remarkable feature of the evidence is that the greatest effects on student learning occur when teachers become learners of their own teaching, and when students become their own teachers. (p. 18)

One way to reach this level of teachers becoming learners and students becoming teachers is to participate in a lesson study. Using the lesson-planning tool we referenced previously, you can collaboratively plan the strategies shared in the resource and then observe to see best practices related to implementation and student engagement. Although a full lesson study requires weeks of collaborative planning and observation, you could use the following lesson-planning steps for just one lesson per unit. A complete lesson study does not have to be conducted for every lesson (see also Kanold, 2015a).

1. The team plans a lesson collaboratively, focusing on student demonstration of learning of a Mathematical Practice. (Teams may choose another form of inquiry related to student learning and engagement as well—it is focused on evidence of student learning.)

2. One teacher teaches the lesson while the remaining team members observe and annotate what *students* are doing and learning during each part of the lesson. You can assign each teacher specific students to observe and record data on throughout the lesson.

3. The team looks at each part of the lesson and determines the task's strengths and weaknesses related to student learning. The team then determines modifications to strengthen the lesson.

4. A second teacher teaches the lesson with the modifications, and the observers again annotate what students are doing and learning during each part of the lesson.

5. The team debriefs and discusses the modifications' pros and cons related to student learning and determines takeaways to put into their own future lessons.

Consider how you and your colleagues can grow your understanding of student learning through intentional lesson design and reflection. If it is too difficult to cover classes for teachers who are observing students during the lesson study, consider videotaping the classrooms and using that for debriefing and deep reflection.

A less formal way to observe students in one another's classes and build a shared understanding of student proficiency with the Mathematical Practices is by implementing a walkthrough activity. You can determine what to look and listen for during a lesson to know whether students are learning self-regulation with the Standards for Mathematical Practice. You can also choose to look at engagement and how active students are in learning mathematics during a collaboratively designed lesson. Walkthroughs provide useful tools and actions to refine approaches with teaching strategies to maximize student learning.

As Hattie (2012) explains:

> Self regulation relates to developing *intentions* to make decisions about learning strategies, awareness of how to evaluate the *effectiveness* of these strategies to attain success in learning, and *consistency* in choosing the best learning strategies across tasks and content areas. (p. 109)

An informal walkthrough requires teachers to be in one another's classroom during mathematics instruction for ten to fifteen minutes. During that time, the observing teacher notes which Mathematical Practices students demonstrated, how students were engaged in learning, or some other relevant instructional practice to discuss later as a team. For example, consider the following questions: What did you see related to your inquiry? How many students demonstrated understanding or proficiency? How might the team refine the lesson in the future to promote more students working, engaged, and demonstrating proficiency of a Mathematical Practice?

Tim Kanold, Diane Briars, and Skip Fennell (2012) also have suggested questions to ask:

> There are a few broad-brush things to look for in any mathematics lesson:
> - Who is doing the mathematical thinking—teachers, students, or both?
> - What is the goal of the instruction—understanding mathematics or simply getting answers?
> - What is the cognitive demand of the tasks students are being asked to do? What happens to the demand as teachers introduce the task, and as students work on the task?
> - To what extent are *all* students engaged in mathematics learning? (p. 73)

The suggested strategies in this book contribute significantly to the lesson-planning process. As you read them, consider your own classroom implementation as well as how to use them as a learning tool for you and your colleagues as you work to improve student engagement in learning mathematics.

Overview of Chapters

In the next eight chapters, you will explore why each Mathematical Practice is critical to student understanding and learn how to develop that habit of mind while teaching. You will find an opening description of the Mathematical Practice with an example of a high-level task. In addition to sample tasks designed to engage students in the specific Mathematical Practice, each chapter also provides the following features.

- **Student Evidence and Teacher Actions:** A table describing students' demonstrations of the Mathematical Practice as well as teacher actions to engage students

- **Understand *Why*:** A section offering specific research and summaries from leading educators on the rationale for the practice

- **Strategies for *How*:** A section providing grade-level tasks and questions to use in your classrooms

- **Lesson Example:** A detailed lesson plan featuring the Standard for Mathematical Practice with content and tasks

In all eight chapters, we offer grade-level tasks. The content of the tasks matches content found in the Common Core State Standards for Mathematics and in the standards of other states as well. See appendix E (page 257) for a table containing the domains and clusters of the Common Core State Standards for Mathematics by grade level. Take the time to consider using these tasks even if the suggested grade level differs from your own. With slight modifications, you can use the same questions to engage your students in the specific Mathematical Practice being featured in the chapter.

Your state may or may not be implementing the content identified in the CCSS, but the importance of students being engaged in mathematical thinking as demonstrated by reasoning, critiquing, using multiple representations, making connections, and exploring patterns remains undeniable. Such engagement deepens students' understanding.

As you explore this resource, we hope that you find opportunities to collaborate with your colleagues and reflect on those strategies that you implement with your students. Although our strategies do not represent an exhaustive list, they do provide a springboard to stimulate your thinking and deepen your understanding of the nature and critical role of the Standards for Mathematical Practice as students are learning mathematical content.

Chapter 1

Standard for Mathematical Practice 1:
Make Sense of Problems and Persevere in Solving Them

It is impossible to overstate the importance of problems in mathematics. It is by means of problems that mathematics develops and actually lifts itself by its own bootstraps. . . . Every new discovery in mathematics results from an attempt to solve some problem.

—HOWARD EVES

When Sarah's oldest son was in third grade and working on homework, he asked, "Mom, can you please help me with this one?" while pointing to a word problem. Naturally, she requested he read the problem aloud. He quickly responded, "Mom, I don't need to *read* it. These are all multiplication word problems. Every problem has two numbers; I just don't know how to multiply *these* two numbers together." Sure enough, after reading the problems posed, Sarah saw that he was right.

It struck her that her son's teacher thought students were reading the word problems to make sense of the problem, but students knew every question had two numbers, always factors, and they simply needed to grab the two numbers (without reading) to multiply them.

How can you really teach students to make sense of problems? How can you build student perseverance? This chapter will focus on strategies designed to develop these sense-making and stamina-inducing habits of mind in students. The selection and implementation of high-level tasks is the critical first step to deepening a student's ability to engage in the first Standard for Mathematical Practice with grade-level content.

Figure 1.1 (page 14) shows a high-level task that simultaneously develops content understanding while building students' abilities to make sense of and persevere in solving problems. It also shows the work of two students who used different strategies to solve each task. Students more effectively make sense of tasks when they know there is more than one solution pathway.

Grade 1

Carl earned 15 tickets to spend at the school store.

The numbers of tickets needed to buy each item are in the chart.

Item	Number of Tickets Needed
Sticker	1
Pencil	3
Bookmark	5

Carl thinks he can buy two of each item. Is he correct? Use numbers, words, or pictures to explain your answer.

Student A   **Student B**

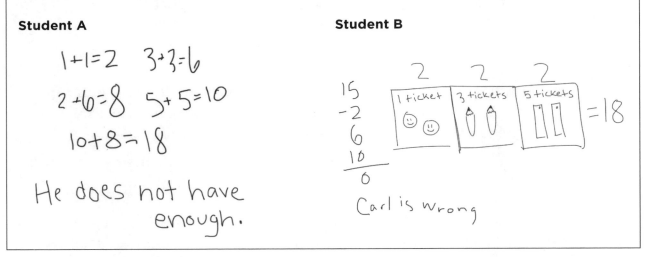

Figure 1.1: High-level task for Mathematical Practice 1.

*Visit **go.solution-tree.com/MathematicsatWork** for a free reproducible version of this figure.*

In this task, student A doubled the tickets needed for each item and then found the sum of the doubled number of tickets. He showed that Carl needed eighteen tickets to buy two of each item. Student B first tried to subtract the doubled number of tickets from fifteen tickets and reached zero before reaching the needed eighteen tickets. He then drew a picture to make sense of the task and summed the number of tickets needed to buy each item and then mentally added 9 + 9 to show he needed eighteen tickets. These two solutions show three different strategies.

The first Standard for Mathematical Practice ("Make sense of problems and persevere in solving them") focuses on students' abilities to understand a task and persevere in solving it using an intentionally chosen strategy while monitoring the reasonableness of the approaching answer along the way. Sometimes problems can have too little information or too much information. In aligning with the content standards, problems must not stop at one-step word problems or a single operation. It is important that tasks build student stamina for problem solving.

Students engage in this Mathematical Practice when they can explain the meaning of a problem in their own words and analyze given information to develop a plan for solving the problem. Students will be able to think of several strategies that they could use to solve the problem and choose one that they feel works most effectively. Additionally, students will evaluate their progress toward a solution and make revisions if needed as well as check the reasonableness of the final solution. Through this work, students will be able to make connections between the strategies others use to solve the problem.

Table 1.1 shows more examples for what students do during a lesson to demonstrate evidence of learning Mathematical Practice 1 and actions you can take to develop this critical thinking in students.

Table 1.1: Student Evidence and Teacher Actions for Mathematical Practice 1

	Student Evidence of Learning the Practice	Teacher Actions to Engage Students
Make sense of problems and persevere in solving them.	Students: • Make a plan to solve a task prior to starting the solution pathway • Articulate how the task given is similar and different from previous tasks • Identify the critical information in the task they need to solve the problem • Make changes in the solution pathway as needed to end with a reasonable answer • Determine more than one strategy to use to solve a task • Estimate an answer prior to solving the task • Evaluate the reasonableness of the solution	Teachers: • Model and connect different strategies that students can use to solve tasks • Ask questions and provide manipulatives that allow for students to productively struggle without lowering the cognitive demand of the task (see Assessing Questions in the sample lesson plan in figure 1.17, pages 36–38) • Have students share solution strategies and make connections • Provide nonroutine, quality tasks for students to make sense of • Encourage students to estimate a solution prior to starting the task • Model different ways to check the reasonableness of an answer • Introduce and connect strategies used in class with those used in other grade levels • Provide opportunities to deepen understanding of students as needed (see Advancing Questions in the sample lesson plan in figure 1.17, pages 36–38)

*Visit **go.solution-tree.com/MathematicsatWork** for a free reproducible version of this table.*

Understand *Why*

Problem solving is central to a mathematics classroom. As early as 1957, George Pólya (1957) wrote that problem solving is not about teaching specific problems but rather the reasoning and critical thinking needed to solve any problem. Making sense of problems requires students to experience novel tasks throughout their units of instruction rather than only procedural word problems they have previously practiced. Problem solving extends beyond the final answer. Students must engage in the process of accessing prior knowledge, using tools, questioning and applying strategies, and discussing the problem with others to work toward possible solutions.

Throughout *Principles to Actions*, NCTM (2014) refers to higher-level tasks as ways for teachers to ask quality questions for formative purposes and for students to productively struggle through varied solution pathways. NCTM (2014) explains, "These tasks encourage reasoning and access to the mathematics through multiple entry points, including the use of different representations and tools, and they foster the solving of problems through varied solution strategies" (p. 17).

Adapting work from Smith (2000), NCTM (2014) defines success as students actively solving tasks without giving up while teachers support student learning without lowering the cognitive demand of the thinking required to solve the task. Students can explain their thinking and the reasonableness of their strategies and answers, question the reasoning of others, and use tools to determine solutions (NCTM, 2014).

In order to use these tasks productively in classroom instruction, tasks teachers choose should be intentional (Kanold, 2012b). Which content knowledge will students demonstrate? Which strategies might students use? How will the questions be structured to support *all* students solving the task without lowering the cognitive demand of the problem? Kanold (2012b) suggests six questions to use when choosing tasks to support this Mathematical Practice.

1. Is the problem interesting to students?

2. Does the problem involve meaningful mathematics?

3. Does the problem provide an opportunity for students to apply and extend mathematics?

4. Is the problem challenging for students?

5. Does the problem support the use of multiple strategies?

6. Will students' interactions with the problem provide real information about students' mathematics understanding? (pp. 28–29)

In *It's TIME: Themes and Imperatives for Mathematics Education*, the National Council of Supervisors of Mathematics (NCSM, 2014) offers two research-affirmed instructional practices that correlate to high levels of student achievement that fit this Mathematical Practice. They are:

Effective teachers of mathematics elicit, value, and celebrate alternative approaches to solving mathematics problems so that students are taught that mathematics is a sense-making process for understanding why and not memorizing the right procedure to get the one right answer. . . .

Effective teachers of mathematics provide multiple representations—for example, models, diagrams, number lines, tables and graphs, as well as symbols—of all mathematical work to support the visualization of skills and concepts. (NCSM, 2014, pp. 30–31)

As for the stamina students need to solve problems in mathematics, the National Mathematics Advisory Panel's (2008) final advisory report states, "Children's goals and beliefs about learning are related to their mathematics performance. . . . When children believe that their efforts to learn make them 'smarter,' they show greater persistence in mathematics learning" (p. xx). Carol Dweck's (2006) research in this area describes the importance of students having a growth mindset over a fixed mindset, which means, in part, they believe their focused effort contributes to their learning.

Similarly, in *Adding It Up: Helping Children Learn Mathematics* (Kilpatrick et al., 2001), the authors conclude that a productive disposition is an important element in mathematical proficiency. This is "the tendency to see sense in mathematics, to perceive it as both useful and worthwhile, to believe that steady effort in learning mathematics pays off, and to see oneself as an effective learner and doer of mathematics" (Kilpatrick et al., 2001, p. 131). This, coupled with Dweck's research, highlights the importance of a student's belief that effort leads to learning and understanding in order to build the perseverance needed to solve problems. To that end, Jo Boaler (2016) further states, "As students are given interesting situations and encouraged to make sense of them, they will see mathematics differently, as not a closed, fixed body of knowledge but an open landscape that they can explore, asking questions and thinking about relationships" (p. 55).

Strategies for *How*

Teachers agree that students need a plan when solving problems but disagree on whether they should introduce problems prior to students learning an efficient set of steps to do specific problems or after establishing some skills and procedures. In fact, there should be a balance. Sometimes teachers can use high-level tasks to introduce topics and other times to reinforce learning of concepts and skills. However, if students are only making sense of problems through the use of previously learned algorithms, the cognitive demand of learning is minimized. Teachers can use the following strategies at various times in a lesson, with an understanding that students may or may not have been introduced to the concepts and skills posed in the task. The consistent theme with each strategy is that *students* are the ones making sense of the task and persevering in solving each one.

Using Graphic Organizers for Building Sense Making

Teachers have long used graphic organizers during literacy as a way for students to make sense of text and organize ideas and facts for writing. They also can help students make sense of mathematics problems and write coherent solutions.

As early as third grade, students learn to write an equation for a word problem using a letter to represent the unknown. Prior to that, they use other symbols such as a blank line, question mark, or open rectangle. Students sometimes struggle with translating the information in text to a mathematical representation. Graphic organizers help students make better sense of the problem.

Part-Part-Whole

For problems involving addition and subtraction, use a part-part-whole model like a tape diagram or a number bond (see figure 1.2).

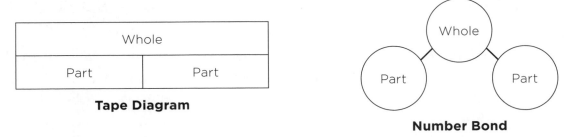

Figure 1.2: Part-part-whole graphic organizers.

Teachers can use part-part-whole graphic organizers for tasks such as the following for grade 1.

> José has 3 stickers. Sandy gives him some more stickers. Altogether, José has 8 stickers. How many stickers did Sandy give José?

This task requires students to solve a subtraction equation or a missing addend equation. They can choose which one to use with the help of a part-part-whole graphic organizer. When such an organizer is introduced and used in the primary grades, it will allow students to recall the strategy and use it later on in the intermediate grades as well.

As a graphic organizer, the size of each bar in the tape diagram does not need to be proportional to the value it represents since students do not yet know the answer. Figure 1.3 shows one possible representation.

8	
3	?

Figure 1.3: Sample tape diagram.

Students might then solve either of the equations $3 + ? = 8$ or $8 - 3 = ?$ using the model as a guide. In later grades, students can use a part-part-whole model to show the solution to the following grade 4 task.

Devon had 6 pairs of socks. He bought 2 packages of socks. Each package had the same number of pairs of socks. Now he has 22 pairs of socks. Write an equation that shows how to find the number of pairs of socks in each package Devon bought.

This task might be represented using the number bond graphic organizer in figure 1.4 using the letter p for the number of pairs of socks in each package.

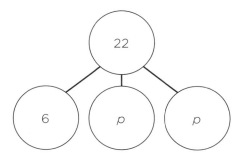

Figure 1.4: Sample number bond.

The student could then write several different equations to show how to find the number of pairs of socks in each package Devon bought. There is more than one way to represent the problem, as the following examples illustrate.

$$22 = 6 + p + p \qquad\qquad 22 - p - p = 6$$
$$22 - 6 = p + p \qquad\qquad 6 + p = 22 - p$$

To find the number of pairs of socks Devon added to his collection, most students will find $22 - 6 = 16$. Then, they will divide the number of pairs of socks purchased in half to find the amount in each package: $16 \div 2 = 8$. There are 8 pairs of socks in each package.

What I Know, and What I Need to Know

When problem solving, students need to determine what information to use from the problem and what information they need to find. Using scaffolded supports helps students identify what they know and what they need to know. Over time, students will begin to ask themselves these very important questions as they read and make sense of problems. Figure 1.5 (page 20) illustrates how a student might use this graphic organizer.

Task
Grade 5 Laura ran $3\frac{2}{3}$ miles each day on Monday, Wednesday, and Friday. On Tuesday and Thursday, she ran $4\frac{1}{2}$ miles each day. On Saturday, she ran a distance so her total distance for Monday, Wednesday, and Friday equaled her total distance for Tuesday, Thursday, and Saturday. How far did Laura run on Saturday?

What I Know	What I Need to Know
Laura ran: $3\frac{2}{3}$ miles Monday $3\frac{2}{3}$ miles Wednesday $3\frac{2}{3}$ miles Friday $4\frac{1}{2}$ miles Tuesday $4\frac{1}{2}$ miles Thursday M + W + F = T + Th + S	The number of miles Laura ran Saturday

Work and Solution
Total miles Laura ran for Monday, Wednesday, and Friday: $3\frac{2}{3} + 3\frac{2}{3} + 3\frac{2}{3} = 11$ miles Total miles Laura ran for Tuesday and Thursday: $4\frac{1}{2} + 4\frac{1}{2} = 9$ miles The total number of miles must be equal so Laura ran 11 – 9 = 2 miles on Saturday. Answer: Laura ran 2 miles on Saturday.

Figure 1.5: What I Know, What I Need to Know template and example.

*Visit **go.solution-tree.com/MathematicsatWork** for a free reproducible version of this figure.*

Frayer Model

The Frayer model is a graphic organizer that Dorothy Frayer and colleagues at the University of Wisconsin created to help students make sense of words (Frayer, Fredrick, & Klausmeier, 1969). In literacy, students place a vocabulary word in the center of the graphic organizer and write its essential and nonessential characteristics as well as provide examples and nonexamples in the surrounding boxes.

In mathematics, the organizer is adapted to have students provide a definition and facts or characteristics as well as examples and nonexamples in the surrounding boxes. The word placed in the center can be a mathematics academic vocabulary word or concept. It's most effective after students have interacted with the topic or word rather than as an introductory activity. Students can also add to the Frayer model as learning continues. Figure 1.6 offers sample student work using the Frayer model to make sense of the word *triangle*.

When using Frayer models, students can fill in models independently or work in groups to complete them. Additionally, you can place a piece of chart paper with a Frayer model at stations throughout the classroom and put a different word at the center of each. Students complete their assigned item and then rotate through the other stations, adding information to each Frayer model along the way. Each group uses a different colored pen or pencil to clearly see which group added information to the models.

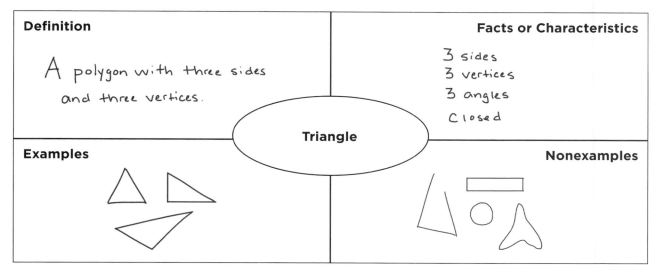

Source: Adapted from Frayer, Fredrick, & Klausmeier, 1969.

Figure 1.6: Sample Frayer model for a triangle.

Visit **go.solution-tree.com/MathematicsatWork** *for a free reproducible version of this figure.*

Reading Word Problems

Key words often help a student know which operation to use when solving a word problem. However, consider the following two tasks, which both use the key words *more* and *altogether.*

Task 1
Trent has 5 toy cars. Alfonso gives him 7 more cars.
How many cars does Trent have altogether?

Task 2
Trent has 5 toy cars. Alfonso gives him some more cars.
Altogether, Trent has 12 cars. How many cars did Alfonso give Trent?

Most students will solve task 1 using *addition*: 5 + 7 = 12. Trent has 12 toy cars. Most students will solve task 2 using *subtraction*: 12 − 5 = 7. Alfonso gave Trent 7 toy cars.

 The key words *more* and *altogether* do not necessarily mean a student should add. If these key words are put on a word wall under *addition*, they must now also be included under *subtraction* given these strategies, and students must be told finding these words is not enough. They must make sense of the entire question as if it were a close-reading question in literacy and locate the information from the text that provides them with the needed information to answer the question posed.

There are some literacy strategies for teaching reading that work for teaching students to read word problems as well. Consider the following grade 3 task for students to read and solve.

> Nadia bought three piñatas for a birthday party. She filled each piñata with 52 candy bars and 38 lollipops. How many total candies did Nadia use to fill all of the piñatas?

Following are three different strategies you might use with students to read and make sense of the task.

Annotations

Readers annotate text when they write the metacognitive thoughts and questions swirling in their minds while reading. Students learn to annotate literary and informational text, and they can apply that same reasoning to mathematical tasks. When annotating, have students take the following five steps.

1. Write their thinking while reading the task.

2. Reread the problem (several times, perhaps) and add to the annotations, if needed.

3. Write what the problem is asking them to find.

4. Share their understanding of the task with a partner.

5. Solve the task with a partner or independently.

A student might make sense of a problem by annotating text like figure 1.7.

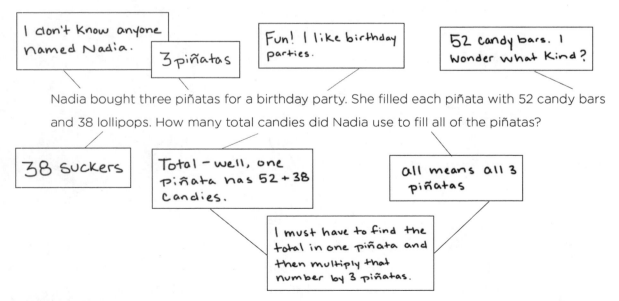

Figure 1.7: Sample annotated text.

Think-Alouds

Think-alouds are similar to annotating text, except a student says out loud what he or she is thinking or wondering about while reading, rather than writing down the thinking. When students read literary or informational text passages, they often alternate who is reading and talking by paragraph. In mathematics, students can alternate by sentence or take a turn reading and thinking through the complete task one at a time.

This strategy is best done with students in pairs—one student reads and stops to explain his or her thinking at the end of each sentence while the other asks clarifying questions. Then, the partners switch and the second person reads and states his or her thinking and questions while the first partner asks clarifying questions. The first narrative might sound like the discussion in figure 1.8.

Student 1 Reads Narrative	Student 1 Thinks Out Loud; Student 2 Listens and Then Asks Clarifying Questions
Nadia bought three piñatas for a birthday party.	**Student 1:** Three piñatas—sounds like a fun birthday party. **Student 2:** What is a piñata? **Student 1:** A thing you hit with a bat until it opens and candy spills out. **Student 2:** Why does someone need three of them? **Student 1:** I don't know.
She filled each piñata with 52 candy bars and 38 lollipops.	**Student 1:** That is a lot of candy. I bet I could scoop it all up. **Student 2:** Would you rather have the candy bars or the lollipops? **Student 1:** The candy bars. All fifty-two of them.
How many total candies did Nadia use to fill all of the piñatas?	**Student 1:** "Total" . . . I think they mean fifty-two plus thirty-eight. **Student 2:** That is for one piñata. What about for all three piñatas? **Student 1:** Oh yeah. We will have to find it three times.

Figure 1.8: Sample think-aloud.

Next, students switch roles and read the task again, creating a deeper understanding of the question to solve. This process is done without a pencil and paper. After each student has had the opportunity to read, then the two students work together to solve the task, discussing their strategies as they work.

One possible variation to this strategy would be to have the second student summarize what the first student read and noticed, rather than ask clarifying questions. Reciprocal teaching is a similar strategy and will be explored further in chapter 6 (page 159).

Some possible prompts and question starters for students to use include the following.

- I wonder why . . . ?

- I agree with what you said because . . .

- I'm not sure that is what the problem is saying. What if . . . ?

- Why did they tell us . . . ? What does it mean? How will we use the information?

- What is this question asking us to find?

- What do we know? What do we need to know?

Highlighting, Underlining, and Circling

Students use highlighters or underlining and circling to mark important information when reading literary and informational text. They can also use these tools when reading mathematical problems. For example, ask students to use a yellow highlighter to highlight the question they are asked to solve and have them use a different color to highlight the important information they will use to solve the problem. If highlighters are unavailable, students can underline the entire question they are going to solve and circle the important information (see figure 1.9).

Nadia bought three piñatas for a birthday party. She filled each piñata with 52 candy bars and 38 lollipops. How many total candies did Nadia use to fill all of the piñatas?

Figure 1.9: Using underlining and circling.

Be consistent throughout the year so students learn a long-term strategy that emphasizes what they need to know and what they need to find. Notice this is not about simply highlighting or circling key words as illustrated in figure 1.9; when teaching this strategy, have students share their highlights or underlines with one another. This can be done in partners, in groups, or with a document camera with the whole group. Have students determine highlights that are helpful and those that are not helpful for solving the problem. Then, provide students with another problem to practice.

Chunking

When students learn to read for comprehension, they must build stamina. Perseverance in mathematical problem solving also requires stamina. Students may not be able to productively struggle through an entire solution and may need to initially have the work chunked into more manageable pieces until they build the stamina to solve the entire task.

The strategies in this chapter help build perseverance through students making sense of the problem. Sometimes, students need help starting the solution. Have students read and start a solution independently. After about three minutes, have them share their initial step and thinking toward the final solution with a partner. Then, have the two students agree on one of the "starts," and finish the problem together using that strategy. This allows students to document some kind of thinking or strategy and begin to question the reasonableness of a strategy prior to taking the time to complete the work and check their answer. Figure 1.10 shows an example of this process.

Task	Nadia bought three piñatas for a birthday party. She filled each piñata with 52 candy bars and 38 lollipops. How many total candies did Nadia use to fill all of the piñatas?		
Student 1	**Student 2**	**Student 3**	**Student 4**
Find 52+38. Multiply the answer by 3.	Find 3×52 and 3×38 and add them together.	Find 52+38. Add the answer three times.	Find 52+38?

Figure 1.10: Possible starts to the piñata task.

Students first make sense of the problem. Suppose students 1 and 2 are paired together as are students 3 and 4. Each pair will have to agree on a strategy and use it to collectively solve the problem.

Student 4 has a good idea for beginning the solution but is not confident and not sure what to do next. This process allows student 4 to be engaged in mathematics and not quit because he or she is unsure of how to begin and also allows students to see that they can use multiple strategies to solve problems.

Students 1 and 2 may determine which strategy to use once they begin the process of simplifying their expressions. In third grade, students often multiply two single digits together and a single digit by a multiple of 10. The strategy student 1 used requires students to simplify $(52 + 38) \times 3$, which is 90×3. Next, students find $(9 \times 3) \times 10 = 27 \times 10 = 270$.

The strategy student 2 used requires students to simplify $(3 \times 52) + (3 \times 38)$. At this point, students may be stuck. Some might think of this as $(3 \times 50 + 3 \times 2) + (3 \times 30 + 3 \times 8)$ or $(150 + 6) + (90 + 24) = 270$.

Rich and meaningful conversations come from students having to agree on a solution pathway and collectively solve the task.

Visualization

For some students, the question's context is not something they are familiar with. Prior to reading the task, you might show pictures or short video clips to let students see or hear the problem's context. For example, if a problem references lacrosse, show a snippet of a game or pictures from a game. If a problem references a farm, show a picture of a farm and talk about one.

When you're unable to show pictures or videos, have students act out the scenario to make sense of the problem before solving it. A few students can present in front of the class or within groups. You might also have students draw their own diagrams or pictures and share them with others.

Understanding Operations

Too many students think mathematics is about finding the one correct set of steps to apply when solving problems. In fact, mathematics requires students to know and be able to apply and reason through multiple strategies so they can choose the most efficient strategy for any problem and understand other students' work. Sometimes learning strategies also prepare students for future mathematics learning in later grade levels and courses. Procedures and fluency always follow conceptual understanding.

Suppose a task asks students to find 12 × 6, and the problem requires knowing how many eggs are in six one-dozen cartons. Figure 1.11 offers some possible student thinking.

Strategy	Mathematics Representation
Memorization	12 × 6 = 72
Repeated Addition	12 + 12 + 12 + 12 + 12 + 12 = 72 OR 6 + 6 + 6 + 6 + 6 + 6 + 6 + 6 + 6 + 6 + 6 + 6 = 72

Distributive Property	(10 + 2) × 6 = (10 × 6) + (2 × 6)
	               = 60 + 12
	               = 72

Doubles	6 × 6 = 36 and 36 × 2 = 72 OR 12 × 3 = 36 and 36 × 2 = 72
Area Model	
Array	= 72
Partial Products	$\quad$ 1 2 $\underline{\times\ \ 6}$ $\quad$ 1 2 ← 6 ones × 2 ones $\underline{+60}$ ← 6 ones × 1 ten $\quad$ 7 2
Standard Algorithm	**1** 12 $\underline{\times 6}$ **72**

Figure 1.11: Student representations for simplifying 12 × 6.

These are only a few of the mathematics strategies students may use. However, too many students think that memorization is their only strategy. Students build perseverance when they are stuck and understand they are responsible for making sense of the problem, whether that means acting out the problem, drawing a picture, guessing and checking, or writing an equation. Giving up is not an option. Through sharing students' strategies, they realize there are many ways to solve a problem.

One way to emphasize using different mathematics strategies is to set up stations in the classroom. At each station, have students solve a mathematics task using a different strategy. They can record their thinking in a graphic organizer or on chart paper. If the station uses objects, you or another adult will need to monitor it more closely to record students who can solve the problem using the objects provided. Students write their answer after the first station they visit and then check their answer using the strategy that matches each part of the graphic organizer at the remaining stations visited (see figure 1.12).

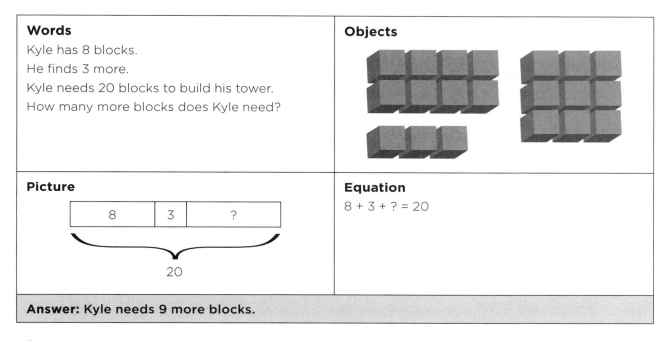

Figure 1.12: Grade 1 strategy graphic organizer example.

*Visit **go.solution-tree.com/MathematicsatWork** for a free reproducible version of this figure.*

For students who understand how to solve problems using various strategies, you can give them a worked-out strategy and have them fill in the remaining boxes in the graphic organizer, including creating a word problem that uses the given strategy and answer.

A few of the strategies students may use to solve addition, subtraction, multiplication, and division problems beyond linking cubes and ten frames are shown in figures 1.13 (page 28) and 1.14 (page 29).

Strategy	Addition Example 13 + 48	Subtraction Example 61 − 13
Base-Ten Pieces	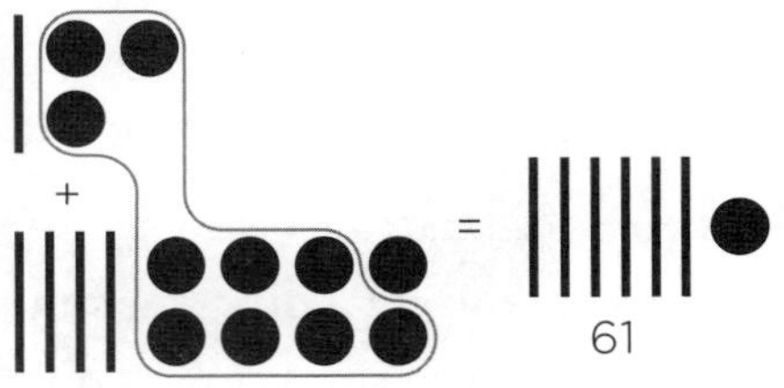 = 61	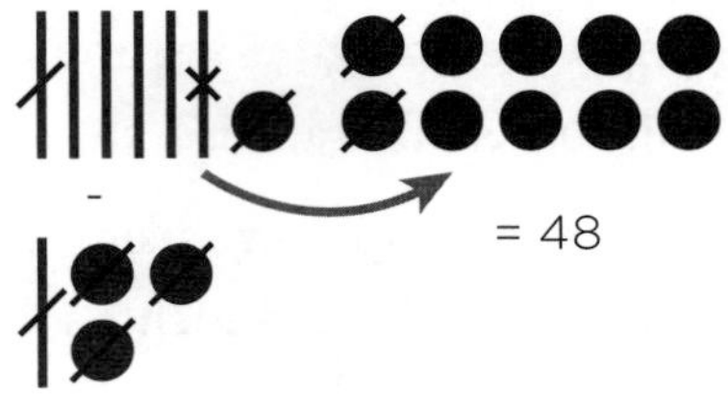 = 48
Open Number Line	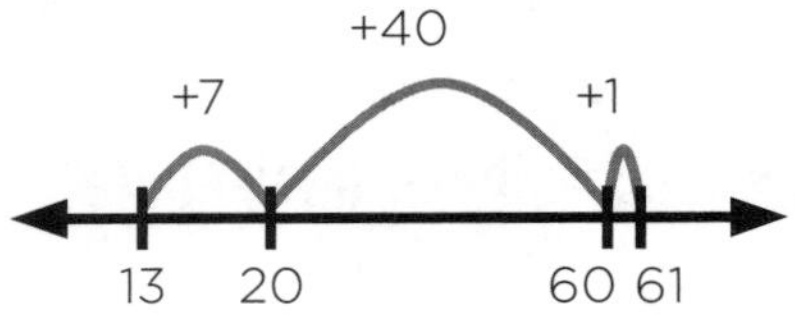 +40 / +7 / +1 / 13 20 60 61	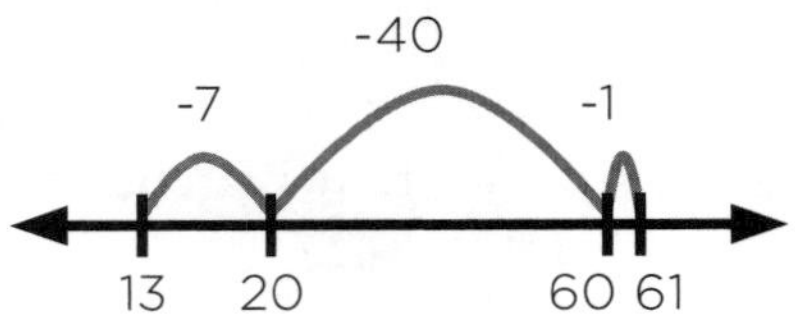 −40 / −7 / −1 / 13 20 60 61
Tape Diagram	? / 13 | 48	61 / 13 | ?
Place Value	13 + 48 / Add the tens. / 50 + 11 / Add the ones. / 60 + 1 = 61	61 − 13 / 61 − 10 = 51 / 51 − 3 = 48
Standard Algorithm	13 / + 48 / 61	5 / 6̸1 / + 1 3 / 4 8

Figure 1.13: Examples of addition and subtraction strategies.

Students learn the standard algorithm after having developed the conceptual understanding of an operation. While the standard algorithm always works, it is not always the most efficient way to solve a problem.

As an example, consider simplifying: $\frac{5}{10} + \frac{6}{12}$.

Students who have only learned the standard algorithm will find a common denominator, rewrite each fraction using equivalent fractions, sum the fractions, and then simplify the result. Yes, this works. However, what if a student read the problem first and then rewrote each fraction as $\frac{1}{2}$ to find $\frac{1}{2} + \frac{1}{2} = 1$? While both strategies "work," the second is more efficient than using the standard algorithm (see Three Es strategy in chapter 4, page 93).

Strategy	Multiplication Example 4 × 256	Division Example 1,024 ÷ 4
Base-Ten Pieces (Groups of)		
	4 × 256 = 8 hundreds + 20 tens + 24 ones = 800 + 200 + 24 = 1,024 1,024 ÷ 4 = 2 hundreds + 5 tens + 6 ones = 200 + 50 + 6 = 256	
Area Model	$$\begin{array}{ccc} 200 & 50 & 6 \end{array}$$ 4 | 800 | 200 | 24 | 4 × 256 = **800 + 200 + 24** = 1,024	$$\begin{array}{ccc} 250 & 5 & 6 \end{array}$$ 4 | 1,000 | 20 | 4 | $4\overline{)1,024}$ = **250 + 5 + 1** = 256
Partial Product and Quotients	$$\begin{array}{r} 256 \\ \times \quad 4 \\ \hline 24 \\ 200 \\ + 800 \\ \hline 1,024 \end{array}$$	$$\begin{array}{r|r} 4\overline{)1,024} & 100 \\ -\ 400 & \\ \hline 624 & 100 \\ -\ 400 & \\ \hline 224 & 50 \\ -\ 200 & \\ \hline 24 & 6 \\ -\ 24 & +\ \ \\ \hline 0 & 256 \end{array}$$
Compatible Numbers and Distributive Property	4 × 256 = (4 × 200) + (4 × 50) + (4 × 6) = 800 + 200 + 24 = 1,024	1,024 = 1,000 + 20 + 4 1,000 ÷ 4 = 250 20 ÷ 4 = 5 4 ÷ 4 = 1 + ____ 256
Standard Algorithm	$$\begin{array}{r} 2\ 2 \\ 2\ 5\ 6 \\ \times \qquad 4 \\ \hline 1,\ 0\ 2\ 4 \end{array}$$	$$\begin{array}{r} 256 \\ 4\overline{)1,024} \\ -\ 8 \\ \hline 22 \\ -\ 20 \\ \hline 24 \\ -\ 24 \\ \hline 0 \end{array}$$

Figure 1.14: Examples of multiplication and division strategies.

Regardless of whether students are making sense of a problem with whole numbers or other rational numbers, have them share their work. Teach them to practice using different strategies to build their logical reasoning as they make connections between the strategies, apply different strategies to verify original solutions, critique the reasoning of others (even those who used unique or different strategies), and try alternate approaches when stuck solving a task.

Estimating Upfront

A teacher asks a student to look at his work and check it after solving a task in class. The student sits at his desk and complies. He looks at his paper but is clearly done thinking about the problem. He waits the appropriate amount of time and returns to his teacher, pronouncing, "I looked at the problem, and I don't see my mistake."

 Too often, teachers ask students to check their work only when they have made a mistake and after completing the problem. In their minds, students have finished the problem and don't understand how to review their work. Using estimation to solve a problem *before* the student picks up a pencil or manipulatives can encourage him or her to work toward that possible answer and change course if necessary. When there is disequilibrium, students are more apt to wonder why and independently work to find the possible error in thinking. This is much like having students make predictions or inferences based on the title of a text or after having read only a portion of the text.

When mathematicians solve problems, they often first predict an answer. What do they expect to happen along the solution pathway? What type of answer makes sense? This then drives some initial thinking about a solution. Students learning to persevere and make sense of problems can do the same.

Before students begin solving a problem, ask:

- "What type of answer makes sense? A number? A name? A color?"
- "What is your prediction for a reasonable answer? Why?"
- "How close was your actual answer to your estimate? Why did they differ? Why were they close?"
- "How did your work confirm or change your estimate?"

After asking for the estimate, collect the possible answers on a piece of chart paper or a whiteboard (someplace visible for students while they work). Follow each estimate with "Why?" to build reasonableness. If applicable, have students agree or disagree with each estimate and discuss why. It is important that you facilitate this discussion, rather than evaluate estimates, so students begin to think this way independently and you can build the classroom culture as a collaborative community of learners.

Once students solve the task, they can identify the closest estimate and furthest estimate. Ask students to explain the thinking that led to a close estimate and the error in thinking that led to an estimate farther from the actual answer. These questions do not need to be asked directly to the students who proposed the original estimates, but rather can be asked of the entire group to consider and brainstorm in partners.

What if none of the estimates are close? Have students prove their answers are correct by showing a second solution strategy, or have students share their solutions with one another as a class to verify the final answer. Then ask, "How can we get closer estimates before doing the problem? What do we need to think about?"

In kindergarten, students can practice estimating the number of objects in a jar each day. They can write their name and number on a sticky note and place it on a whiteboard or designated wall space. Students will, at first, count the objects in the jar. Work to have students estimate without counting first, and then count the objects in the jar afterward as a class. Talk about which numbers are less than, equal to, or greater than the actual value of objects in the jar and how close the estimates are to build an understanding of number and number comparisons.

Problem Solving

Problem solving is an art and a skill students develop as they engage in Mathematical Practice 1. Following are a few specific ideas for how to support students as they learn to solve problems.

Problem-Solving Plan

Many heuristic methods teachers and students use to frame quality problem solving often stem from Pólya's (1957) four principles in *How to Solve It*.

1. Understand the problem.
2. Devise a plan.
3. Carry out the plan.
4. Look back.

Students must first make sense of the problem when presented with a task to solve. Previously in this chapter, we discussed that students need to identify the question and the information they need to solve the problem. They might also think about other problems they have solved similar to the one posed, or maybe they need to read and reread the problem.

Next, students must plan their solution pathway. Some possible problem-solving strategies students might use include the following.

- Guess and check
- Draw a picture
- Use a model

- Look for a pattern
- Make a list
- Make a table
- Make a graph
- Work backward
- Use a formula
- Create and solve an equation
- Solve a simpler problem

Once they've devised a plan, students must persevere in carrying out the plan and revising along the way as needed. Finally, they must check the reasonableness of their answer in the context of the original problem. This might include solving the problem using a second strategy to verify the original solution. Does the answer make sense? Does it answer the question posed?

Consider a consistent sequence of questions to ask students when they are problem solving. Using a sequence of questions allows students to begin to ask themselves the same questions as independent problem solvers. Ideally, teachers will ask these throughout the year and consistently ask them throughout grades K–5 so students see connections when solving problems. Following is an example series of four questions.

1. What is the problem asking?
2. How will I solve it?
3. Is this strategy working?
4. Does my answer make sense?

Teachers or students can pose these questions before, during, and after the problem-solving process.

Groups and Roles

Meaningful and purposeful dialogue improves student learning. In fact, John Hattie's (2009) book *Visible Learning* cites Alexander's (2008) research regarding the benefits to learning when students are the ones asking questions and responding to ideas, rather than all authority coming from the teacher. Teachers can learn from listening to students, and students learn from understanding and building on others' ideas. If, however, you ask students to work in groups, the activity or task must require interdependence to reach the solution.

 One way to have students make sense of problems and persevere in solving them is by instructing them to work together in groups and to interact and talk with one another to productively struggle through an agreed-on solution pathway. This approach also allows for students to practice their speaking and listening skills necessary for proficiency in English language arts. You may need to provide sentence frames, which are discussed as a strategy in chapter 3 (page 63).

At times, you can randomly assign students to groups, and other times, you'll need to intentionally plan groups in advance based on the task and the need for differentiation. Students working in groups need to know that they must rely on each other as resources prior to you answering any questions. Ideally, students in the groups will have assigned roles. Table 1.2 lists possible roles to consider for groups of four students but is not an exhaustive list. Note that a reporter is omitted since every student should be able to report on the solution pathway and final answer.

Table 1.2: Student Roles During Group Work

Role	Description
Facilitator or Planner	This student moves the group forward in its work and keeps conversations on task. He or she reads the problem to the group and makes sure every student is contributing to the solution. The facilitator or planner keeps track of time.
Questioner	This student asks the group clarifying questions when making sense of the problem and planning and executing a strategy. At the end, he or she always asks the question, "Does our answer make sense?" This student also keeps track of questions the group asks and makes sure they all get answered. He or she makes sure students identify what they know and what they need to know to solve the problem.
Recorder	This student records the group's thinking and draws pictures or writes equations to discuss and use in the solution pathway.
Resource Manager	This student acts as the liaison between the group and the teacher. When the entire group is stuck, the resource manager asks the teacher or another group a question on behalf of the group and has the responsibility of sharing the answer with the group upon returning. The resource manager also gathers any necessary supplies.

Cards are useful tools to remind students of their roles. Figure 1.15 (page 34) shows sample cards to use with the roles in table 1.2.

Even with roles, students must be held accountable for their solution pathway and interdependence. This might mean each student writes down the solution on his or her own paper or in a journal. It might involve another group strategy like Four Corners (discussed in the next section). All students must be able to explain their strategy and solution. Collaborative learning is most effective when teams build interdependence and all members are held accountable for their learning.

Facilitator or Planner

Make sure all students are participating.

Read the task.

Watch the time and make sure the group is on task.

Questions to ask the group:

- Who has another idea?
- What questions do you have about our work?
- Did we answer the question?

Questioner

Work with the group to make sure everyone understands the task and the solution plan and can explain whether the answer makes sense.

Questions to ask the group:

- What do we know?
- What do we need to know?
- What is our plan?
- Did we all get the same answer?
- Does our answer make sense?

Recorder

Show the thinking of the group that leads to the solution.

Draw a picture.

Write an equation.

Questions to ask the group:

- Is this chart correct?
- Is this what you meant?
- How could we show this another way?

Resource Manager

Access help from the teacher or another group when the entire group is stuck and then share the answer with the group.

Questions to ask the group:

- What do you think we could try?
- Who can explain what we have done so far?
- Where are we stuck?
- What are we missing?
- What other information do we need?
- Does this answer make sense to everyone?

Figure 1.15: Sample role cards.

*Visit **go.solution-tree.com/MathematicsatWork** for a free reproducible version of this figure.*

Four Corners

Four Corners is an activity for groups of four. Students sit around a large piece of paper (about half the size of a piece of chart paper). They fold the paper or draw lines to separate it into four quadrants. They draw a circle or rectangle in the center of the paper like figure 1.16.

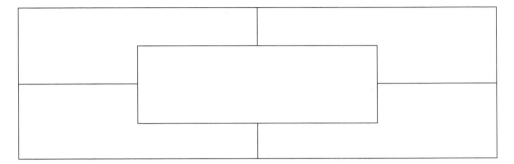

Figure 1.16: Four Corners diagram.

The teacher gives each group a higher-level task to solve. Often, groups are given the same task, though sometimes the teacher may want each group to solve a unique task to then share with the class. For example, a possible task to use for kindergarten might be as follows: Charlie has 14 apples. Margaret has 1 less than 16 apples. Who has more apples? Show how you know your answer is correct.

Each student spends time writing a possible solution in his or her quadrant. This helps hold all students accountable for making sense of the problem and starting a solution pathway. The solutions will not all be pointing the same way since students are facing different directions when they write.

After some independent thinking and writing time (about five minutes), have students share their initial thinking. If any student is stuck, this may provide him or her with an opportunity to solve the task. Provide another five minutes of independent work time.

Finally, have students share their solutions with one another and then determine which solution they want to have as their final solution representing the group's work. They can orally share their solution, or students can rotate the chart paper, read the solution, and then ask clarifying questions as needed. Students write the final solution in the inside circle or rectangle, or you can choose to not have students initially draw the rectangle on their papers and simply use a colored marker to draw a border around the solution to share.

Later, during the mathematics block of time or the following day, hang the posters on the wall. Have students read others' work so they can discuss the various strategies peers used.

 ## Talk-Record

For this activity, students are in pairs, and one student has the pencil. Like reciprocal teaching in literacy, one student will teach the other and then reverse roles to better understand the problem. The following example illustrates the structure and content of a pair's conversation.

Student 1: Reads the problem aloud and identifies what he or she knows and what he or she needs to know

Student 2 (with pencil): Records the thinking of student 1

Student 2: Determines aloud a strategy to solve the problem

Student 1 (with pencil): Records the strategy and work from student 2

Student 1: Determines if the answer makes sense

Student 2 (with pencil): Records the justification for why the answer makes sense

There are many variations of the talk-record strategy. For example, one student might do the recording, and then the roles are reversed. The second student then verifies the thinking while the first records any additional information needed. Another variation is for one student to solve the task using a strategy the second student records; then the second student articulates a second strategy to solve the problem that the first records, and the two make connections between the strategies and final solution. Ultimately, one student is speaking, while the other is listening, recording, and asking questions as needed. Together, they make sense of the problem and persevere in solving it.

Lesson Example for Mathematical Practice 1: Fraction Word Problems

The lesson plan in figure 1.17 focuses on students making sense of problems and persevering in solving them. The tasks included allow for teachers to use several strategies, and students will demonstrate an understanding of fraction word problems based on how they solve the problems. This lesson is designed for fourth-grade classrooms but could also be used with fifth-grade students. Although Mathematical Practice 1 is the focus of the lesson, students will also be talking and working together to justify their solutions and critique the reasoning of others (Mathematical Practice 3), attending to precision through their writing (Mathematical Practice 6), and reasoning abstractly and quantitatively (Mathematical Practice 2). A commentary follows the lesson providing more information related to the rationale and importance of each lesson component. Figures 1.18 and 1.19 (page 39) support the lesson's tasks.

Unit: Fraction Operations (4.NF.4)

Date: March 2

Lesson: Problem solving using fraction reasoning and operations

---

Learning objective: As a result of class today, students will be able to read and solve word problems involving fractions.

Essential Standard for Mathematical Practice: As a result of class today, students will be able to demonstrate greater proficiency in which Standard for Mathematical Practice?

Mathematical Practice 1: "Make sense of problems and persevere in solving them."

- Students will read and make sense of word problems.
- Students will identify key information needed to solve problems.
- Students will persevere to solve a task in a group.

Formative assessment process: How will students be expected to demonstrate mastery of the learning objective during in-class checks for understanding teacher feedback and student action on that feedback?

- Students will share different ways to find $\frac{2}{3} \times 30$, and the class and teacher will make connections between the strategies.
- Students will work in groups to identify which of two tasks has too much information and which is missing information. They will identify the key information they need to solve each task. The teacher will monitor student conversations and check student thinking.
- Students will work in groups to solve a task. The teacher will monitor progress and ask assessing and advancing questions as needed.

Probing Questions for Differentiation on Mathematical Tasks

Assessing Questions	Advancing Questions
(Create questions to scaffold instruction for students who are stuck during the lesson or the lesson tasks.)	(Create questions to further learning for students who are ready to advance beyond the learning standard.)
<ul><li>How can you use a manipulative or drawing to make sense of the problem?</li><li>What else can you try?</li><li>What do you know?</li><li>What do you need to know?</li><li>How will you know if your strategy works?</li></ul>	<ul><li>Write a word problem that requires a student to compute two different operations and has too much information included.</li><li>Write a word problem that requires a student to solve using two steps, includes at least one fraction, and has an answer between 62 and 68.</li></ul>

Tasks (Tasks can vary from lesson to lesson.)	What Will the Teacher Be Doing? (How will the teacher present and then monitor student response to the task?)	What Will Students Be Doing? (How will students be actively engaged in each part of the lesson?)
Beginning-of-Class Routines How does the warm-up activity connect to students' prior knowledge, or how is it based on analysis of homework?	Teacher makes sure all students have a whiteboard and marker at the start of class and have the following question written at the front of the room. Multiply $\frac{2}{3} \times 30$. Show your answer is correct using at least two different strategies. Teacher asks students to share their solutions with the class as shown on whiteboards to generate a list of possible ways to multiply $\frac{2}{3} \times 30$.	Students write their two strategies on the whiteboard to prove $\frac{2}{3} \times 30 = 20$. Students share their strategies with an elbow partner. Students share their work on the whiteboard with the class to generate different ways to think about multiplying $\frac{2}{3} \times 30$.

continued →

Task 1		
Task 1 How will students be engaged in understanding the learning objective? (See figure 1.18.)	Teacher gives pairs of students two cards as in task 1 (figure 1.18) and the directions: • Each person reads one task. • Each person restates the question posed and identifies *What do I know?* • The pair determines which task has too much information and which does not have enough. • The pair determines what else it needs for the task without enough information. • The pair sees if it agrees on the information read in each task. Teacher facilitates a whole-group discussion about strategies used to read tasks and identify the important information. How did students determine what else they needed to know for the problem without enough information?	Each student in the pair reads one card independently. He or she identifies *What do I know? What do I need to know?* Then, he or she determines if the task has too much information or not enough information. The two students determine which task has too much information and which one has not enough information. For the one without enough information, they determine what else they need. The two pairs within a group of four check to see if they came to the same conclusion regarding the two tasks. Students engage in a whole-class discussion explaining their thinking related to reading the task and identifying important information.
Task 2 How will the task develop student sense making and reasoning? How will the task require student conjectures and communication? (See figure 1.19.)	Teacher gives each group of four a large sheet of paper for the Four Corners activity embedded in task 2 (see figure 1.19). Teacher asks groups to make an estimate or prediction for the number of cups Juliette and her mom picked. Teacher monitors student work and asks assessing and advancing questions as needed.	Students work independently to start a solution or solve the task. Students share ideas and work to finish their solution pathway. Students discuss which explanation will be put in the center of the poster to share with the class. Students look at the solutions classmates post and make connections to their own solution.
Closure How will student questions and reflections be elicited in the summary of the lesson? How will students' understanding of the learning objective be determined?	Teacher poses the following questions: • Why should we estimate before solving a problem? • What strategies did you use to read the problem? • What strategies did you use to solve and check the problem? • How did you use fraction understanding to answer the tasks?	Students answer their questions in a journal or as a pair-share and articulate what they learned during the lesson.

Source: Template adapted from Kanold, 2012c. Used with permission.

Figure 1.17: Grade 4 lesson-planning tool for Mathematical Practice 1.

*Visit **go.solution-tree.com/MathematicsatWork** for a free reproducible version of this figure.*

Task 1

Name: _____   Partner's name: _____

Read each problem. Underline what the problem is asking you to find. Circle the important information in the problem you need to use to answer the question.

One problem has too much information in it and the other does not have enough information. Which is which? Check whether you think each problem has too much information or not enough information. If the problem does not have enough information, write what else you need to know to solve the problem.

Problem A	**Problem B**
A jar of jelly beans is filled with red, yellow, and white jelly beans. In the jar, $\frac{1}{10}$ of the jelly beans are red, $\frac{7}{10}$ of the jelly beans are yellow, and $\frac{2}{10}$ of the jelly beans are white. There are 300 jelly beans in the jar. How many of the jelly beans are yellow?	Billy is playing with blocks. He has 16 yellow blocks, 20 red blocks, 50 green blocks, and some blue blocks. What fraction of the blocks are red?

❑ Too much information	❑ Not enough information	❑ Too much information	❑ Not enough information

For problem _____, we also need to know _____

_____

_____

_____.

Figure 1.18: Task 1 for Mathematical Practice 1 grade 4 lesson.

*Visit **go.solution-tree.com/MathematicsatWork** for a free reproducible version of this figure.*

Task 2

Juliette and her parents picked blueberries for 2 hours. Altogether, they picked 50 cups of blueberries. Juliette picked $\frac{2}{10}$ of the blueberries, her mom picked $\frac{6}{10}$ of the blueberries, and her dad picked 10 cups. How many cups did Juliette pick? How many cups did her mom pick? Explain how you know your answer is correct.

Figure 1.19: Task 2 for Mathematical Practice 1 grade 4 lesson.

*Visit **go.solution-tree.com/MathematicsatWork** for a free reproducible version of this figure.*

During the introduction, students work together to remember how to multiply a fraction by a whole number. Given an expression, they will share different ways to prove the product using strategies that may involve manipulatives, number lines, models, algorithms, and so on. Among other strategies, some students will think of $\frac{2}{3} \times 30$ as $\frac{1}{3} \times 2 \times 30 = \frac{1}{3} \times 60 = \frac{60}{3} = 20$. Others will make an area model, such as the one in figure 1.20.

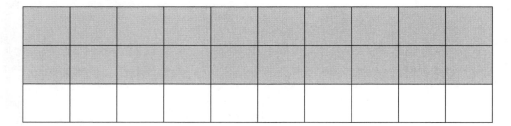

Figure 1.20: Sample area model.

In this model, $\frac{2}{3}$ of the thirty unit squares creating the area are shaded to reveal an answer of 20.

The first task supports students reading a problem as they identify answers to the questions What do I know? and What do I need to know? One of the tasks has too much information, and the other has not enough information. The goal is for students to make sense of the task by identifying which is which and identifying the information they need to solve each task. Students do not need to actually solve the task—this is chunked into only understanding what they need. These tasks could be given as the independent practice (homework) after the lesson.

In task 1 (figure 1.18), problem A has too much information, and problem B has not enough. For problem B, if students ask for the number of blue blocks, tell them fourteen, and if they ask for the total number of blocks, say one hundred. A common mistake on this problem is to sum the blocks without blue and then find the fraction that represents red.

The second task (figure 1.19) provides an opportunity for students to solve a problem using multiple approaches and strategies. The Four Corners strategy will allow students to generate different solution pathways and discuss those that are within one group as well as those demonstrated by the class. You may also want to use some roles for students in the group. Solving this task will require perseverance.

Begin task 2 by giving each student a strip of paper with the question and giving the group the larger paper to write on collectively. As a way to promote verification once the task is complete, have groups generate an estimate or prediction for what they believe the answer may be close to. Then, students will work toward a solution. Two possible solution pathways follow.

1. A student solves $\frac{2}{10} \times 50 = 10$ to find the number of cups of blueberries Juliette picked and $\frac{6}{10} \times 50 = 30$ to find the number of cups her mom picked (or finds $50 - 10 - 10 = 30$).

2. A student determines $\frac{2}{10} + \frac{6}{10} = \frac{8}{10}$. This means Juliette's dad picked $1 - \frac{8}{10} = \frac{2}{10}$ of the blueberries, the same fraction that Juliette picked. Thus, Juliette must have picked ten cups also and her mom picked $50 - 10 - 10 = 30$ cups.

Students may show their computations using different strategies for multiplication or addition and subtraction with fractions too.

One common mistake students make is to calculate that 50 − 10 = 40 cups to find the number of cups Juliette and her mom picked together. Then, students find two-tenths of 40 and six-tenths of 40, instead of 50.

To close the lesson, students will revisit their learning and discuss answers to these three questions.

1. Why should we estimate before solving a problem?

2. What strategies did you use to read the problem?

3. What strategies did you use to solve and check the problem?

You may also want to include a question related to the content of fraction equivalence and operations with fractions. Emphasize the process of making sense of the problem just as much as the content students need to solve the problem.

Summary and Action

Mathematical Practice 1 requires students to learn ways to make sense of problems and build the stamina needed to reach a solution. It will not be satisfactory for students to say "I am stuck and can't do it." Mathematical Practice 1 builds the expectation from a growth mindset that students are responsible for trying a different approach or rereading the problem so that students will instead say, "I am stuck, so now I will try [drawing a picture, writing an equation, and so on]." They will develop problem-solving strategies and metacognitive questions to guide them from the original question to an estimate to an executed plan, leading to a justified solution.

Identify a content standard you are having students learn currently or in the near future. Choose at least two of the Mathematical Practice 1 strategies from the following list to develop the habits of mind in students in order to make sense of problems and persevere in solving them.

- Using graphic organizers for building sense making
  - Part-part-whole
  - What I Know, and What I Need to Know
  - Frayer model
- Reading word problems
  - Annotations
  - Think-alouds
  - Highlighting, underlining, and circling
  - Chunking
  - Visualization
- Understanding operations
- Estimating upfront

- Problem solving
  - Problem-solving plan
  - Groups and roles
  - Four Corners
  - Talk-record

Record these in the reproducible "Strategies for Mathematical Practice 1: Make Sense of Problems and Persevere in Solving Them." (Visit **go.solution-tree.com/MathematicsatWork** to download this free reproducible.) How were all students engaged when using the strategy? What was the impact on student learning? How do you know?

Chapter 2

Standard for Mathematical Practice 2: Reason Abstractly and Quantitatively

Creating classroom opportunities for developing higher-order thinking is essential for helping students become critical thinkers, problem solvers, innovators, and change makers upon which our society thrives.

—PÉRSIDA HIMMELE & WILLIAM HIMMELE

Kit's brother Mike shared the following incident when his fourth-grade daughter came home from school with questions on her division homework. Specifically, she was working on 456 divided by 6. Kit's brother's conversation with his daughter went something like the following.

Mike: Tell me what you know.

Daughter: Well, my teacher said that I could break apart 456.

Mike: What do you mean, "break apart"?

Daughter: I can break it into smaller numbers that make 456.

Mike: This problem is division. Are you dividing when you break numbers apart?

Daughter: I think so, but I can't remember.

Mike: Well, I'll show you how I do it . . .

Daughter: Oh, that's OK, Dad.

Apparently, Mike's process did not make any sense to his daughter, and her strategy did not make sense to him.

The key to building mathematical proficiency hinges on this sense making. When students make sense of the context, use their understanding to solve problems, and value their strategies as useful, they increase their belief that mathematics is a fundamental tool in daily life. Learning mathematics is not about memorizing steps and procedures. Mathematics is about reasoning and making sense of the context and the values presented. The second Standard for Mathematical Practice, "Reason abstractly and quantitatively," emphasizes the critical importance of students' thinking. It focuses on students' abilities to understand quantities in the specific context and operate on these quantities in order to answer the question at hand. Students engage in this Mathematical Practice as they create a logical representation of the problem. Students *decontextualize* as they represent the situation with pictures, objects, or symbolically. After using the representations, students place the results back in the situation to determine whether or not the work makes sense in the given context. *Contextualizing* represents an important characteristic of this Mathematical Practice. Mathematically proficient students in this practice understand the meaning of the quantities—not just how to operate on them.

Figure 2.1 shows a second-grade problem and sample student work.

Grade 2

Alan picked apples and placed them in his basket. In the morning, he picked 22 apples. In the afternoon, he picked more apples. When he arrived home, he counted 56 apples in his basket. Did Alan pick more apples in the morning or in the afternoon?

Explain how you know.

Miller's Work: *Grafton's Work:*

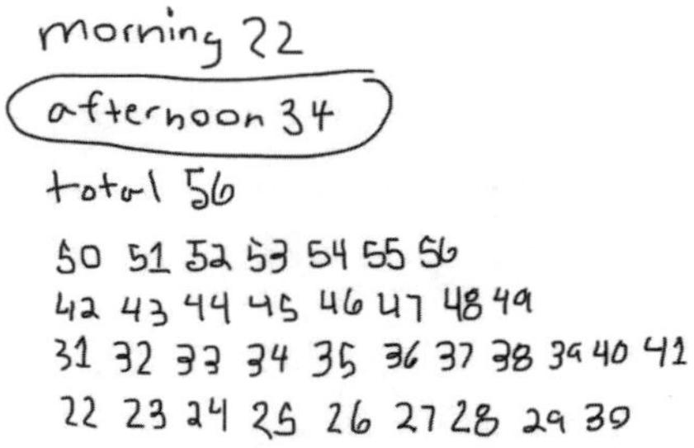

Afternoon

morning 22
(afternoon 34)
total 56

50 51 52 53 54 55 56
42 43 44 45 46 47 48 49
31 32 33 34 35 36 37 38 39 40 41
22 23 24 25 26 27 28 29 30

Figure 2.1: Student work for grade 2 apple task.

*Visit **go.solution-tree.com/MathematicsatWork** for a free reproducible version of this figure.*

Miller explained his thinking orally. He reasoned that 22 is close to 20, and he knew that 20 + 30 = 50. Miller simply answered the question by stating "afternoon," as that was all he needed. Grafton took a different approach. He started at 56 and wrote all of the numbers in descending order until he reached 22. Grafton then counted the numbers he listed to obtain the total number of apples picked in the afternoon. When asked how he started counting, Grafton said, "This is where I'm starting [he points to the number 56]. When I count 1, I am on 55." Both boys arrived at the same correct answer; their reasoning differed dramatically.

Table 2.1 shows more examples for what students do during a lesson to demonstrate evidence of learning Mathematical Practice 2 and what actions you can take to develop this critical thinking in students.

Table 2.1: Student Evidence and Teacher Actions for Mathematical Practice 2

	Student Evidence of Learning the Practice	Teacher Actions to Engage Students

Reason abstractly and quantitatively.	Students:
	• Use multiple representations
	• Identify meaning of quantities within the problem
	• Look for relationships among the values presented
	• Use the context of the problem
	• Make connections between concepts
	• Use mathematical properties
	• Understand the meaning of a number when it contains units
	• Check to see if the answer makes sense
		Teachers:
		• Provide opportunities for students to solve problems
		• Encourage a variety of solution strategies
		• Use understandable contexts for problems
		• Ask questions such as:
		  • "What do the numbers represent?"
		  • "What do you know about this problem?"
		  • "How are the quantities related?"
		  • "What strategy have you used so far? What did you learn?"
		  • "How can you represent this situation with numbers and symbols?"
		• Use think-alouds to model reasoning for those students who are stuck

*Visit **go.solution-tree.com/MathematicsatWork** for a free reproducible version of this table.*

Understand *Why*

In *Principles to Actions* (NCTM, 2014) the second of eight Mathematics Teaching Practices is "Implement tasks that promote reasoning and problem solving," which connects directly with Mathematical Practice 2, "Reason abstractly and quantitatively." As mentioned previously, NCTM (2014) further notes that "effective teaching of mathematics engages students in solving and discussing tasks that promote mathematical reasoning and problem solving and allow multiple entry points and varied solution strategies" (p. 10).

Engaging students in the act of thinking serves two purposes. First, it enables students to make sense of the context. Learners can apply what they understand and pursue their thinking toward a solution. This final solution may or may not make sense given the context, but in the process, students' understanding progresses. They can then determine next steps based on what they have already done. Second, by focusing on students' thinking, teachers can determine the necessary next steps for individual students. By highlighting reasoning, students are active participants in the task at hand, and teachers can use students' demonstration of reasoning for diagnostic purposes.

In *Adding It Up: Helping Children Learn Mathematics* (Kilpatrick et al., 2001), the authors define mathematical proficiency as consisting of five strands: (1) conceptual understanding, (2) procedural fluency, (3) strategic competence, (4) adaptive reasoning, and (5) productive disposition. The authors' explanation of the adaptive reasoning strand underscores the importance of Mathematical Practice 2: "In mathematics, adaptive reasoning is the glue that holds everything together, the lodestar that guides learning" (Kilpatrick et al., 2001, p. 129). The authors also state that children as young as four and five demonstrate evidence of the ability to encode and make inferences: "With the help of representation-building experiences, children can demonstrate sophisticated reasoning abilities" (Kilpatrick et al., 2001, p. 129). In order for reasoning to occur, research suggests that three conditions must be in place: "[Students] have a sufficient knowledge base, the task is understandable and motivating, and the context is familiar and comfortable" (Kilpatrick et al., 2001, p. 130).

Strategies for *How*

Teaching students to reason abstractly and quantitatively is not always easy. It requires purposeful lesson planning and repetition of strategies to develop that reasoning. The strategies that follow provide examples of how to craft tasks and activities to build students' quantitative and abstract reasoning skills.

Quantity Questions

Engaging students in the act of reasoning can be accomplished through intentional planning. In the following grade 4 task, students must go beyond simply comparing the numbers to answer the question. They must read carefully and focus on the value of the number based on the units attached to it.

> The maple trees on Josh Fitz's farm produced 8 gallons of sap in one day. The trees on his neighbor's farm produced 28 quarts of sap during the same day. Whose trees produced more sap on this day? Explain or show how you know.

A fourth grader described her process to her partner after determining the number of quarts in a gallon. She said, "I divided 28 by 4 and then knew that the neighbor only had 7 gallons. I now know that the trees on the first farm made more sap." Her partner said, "I did it starting with the gallons and multiplied them by 4. Then I saw that there were 32 quarts from trees on the first farm." Both students solved the question correctly and listened to each other's approach. They deepened their understanding of the connection between multiplication and division as their teacher asked them to consider how they both got the correct answer using two different operations.

In the primary grades, students in kindergarten and first grade learn that 2 < 5. Then, in second grade they are asked which is longer: 2 feet or 5 inches? This is very strange and abstract for students. If 2 < 5, why is 2 feet > 5 inches? Why do words change the comparative values? In third grade, the idea that $\frac{1}{2}$ can represent half of the people at a table versus half of the students in a school is also very abstract.

To make students aware of the meaning of numbers, teachers must ask them continually to compare numbers with units and must have students always use units in their answers and dialogue for clarity. Following is a short list of example quantity questions.

- Which is greater: 65 cents or 3 quarters?

- Which is heavier: 2 pounds or 30 ounces?

- Which is longer: 10,560 feet or 1.5 miles?

- How can $\frac{1}{3}$ of a pizza be more than $\frac{1}{2}$ of a pizza?

 ## Graphic Organizers for Reasoning About Quantities

In Mathematical Practice 1, you explored graphic organizers students could use to make sense of problems and begin a focused solution pathway. Teachers can also use graphic organizers to help students reason about the meaning of quantities.

Venn Diagrams

First introduced in 1880 by John Venn, a Venn diagram serves to organize students' thinking in terms of similarities and differences (Lucidchart, 2013). In literacy, students use the circles to compare characters in a story or identify similarities or differences between various vocabulary words. Figure 2.2 offers an example of how to use a Venn diagram in mathematics, as students identify numbers that are even and prime. The third question provides an opportunity for students to confront a common misconception that all prime numbers are odd or all odd numbers are prime.

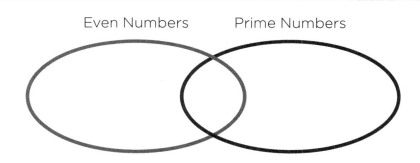

1. Place the following values in the Venn diagram: 12, 18, 23, 29, 47.

2. Where should the number 2 be placed? Explain your thinking.

3. Renee determined that odd numbers must also be prime numbers. Do you agree with Renee? Use words, numbers, or pictures to justify your answer.

Figure 2.2: Venn diagram example.

*Visit **go.solution-tree.com/MathematicsatWork** for a free reproducible version of this figure.*

Figure 2.3 offers another Venn diagram. This time, students place the multiples of 4 in one circle of the Venn diagram and multiples of 6 in the other circle. Students can consider the relationships between quantities in the circles and draw conclusions.

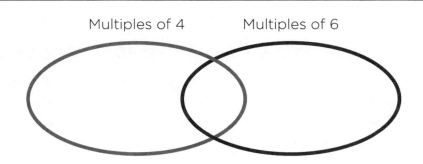

Multiples of 4   Multiples of 6

1. Write the multiples of 4 and 6 up to 60 in the Venn diagram.

2. Explain why you placed numbers in the intersection of the two circles.

3. Would 192 be placed in the intersection of the two circles? Explain how you know your answer is correct.

Figure 2.3: Venn diagram showing multiples of 4 and 6.

*Visit **go.solution-tree.com/MathematicsatWork** for a free reproducible version of this figure.*

Venn diagrams also serve as a data-gathering tool. Teachers can use the results to create mathematical questions based on student data.

Of course, Venn diagrams can include three circles along with an initial question, such as "In my free time, I enjoy . . ." The circles might include such labels as Reading, Going to Movies, and Playing Sports. As these may present hard choices, teachers can ask students to make a decision as to their *favorite* activities. Blocking out some of the intersections indicates that students may not select that intersection (see figure 2.4). In this case, students cannot say that they enjoy all three activities. Students can then use their data to make comparisons such as how many more students enjoy playing sports than going to movies. Students might also make predictions based on this data as to how many students might prefer going to an amusement park as compared to going to see a play. They make mathematical conclusions using the language and structure of the Venn diagram to reason abstractly and quantitatively.

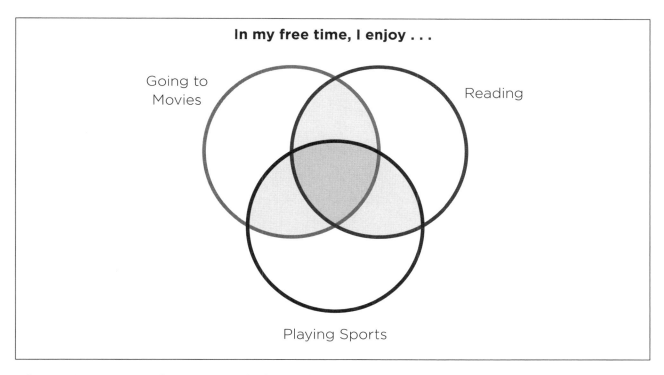

Figure 2.4: Venn diagram with forced choice.

Consider using Venn diagrams as organizational tools that enable students to compare and contrast quantities, which develops a student's ability to reason abstractly and quantitatively (Mathematical Practice 2).

Target Circles

Target circles, another graphic organizer to promote reasoning about quantities, encourages students to consider multiple ways to compose numbers. The target number is inside a circle, and students fill in the rectangles with different number combinations to make the target value. Figures 2.5 and 2.6 (page 50) illustrate two kindergarten versions of target circles.

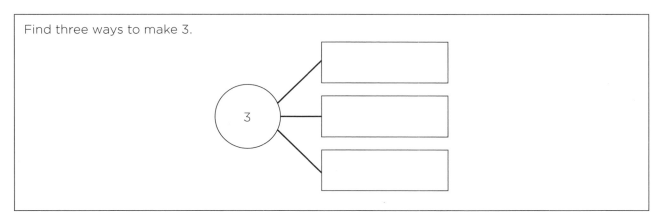

Figure 2.5: Target circles—number combinations for 3.

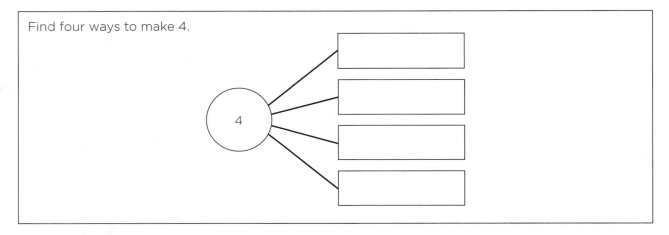

Find four ways to make 4.

Figure 2.6: Target circles—number combinations for 4.

You can stimulate students' thinking by asking them to determine if they have found all of the ways to make 4. This type of open question allows students to demonstrate their understanding of different operations or number relationships.

Target circles are also useful with students in other elementary grades. Figure 2.7 shows an example focusing on place value appropriate to grades 1 or 2. Students will be reasoning abstractly and quantitatively, as they understand there is more than one combination of tens of ones that they could use to describe a two-digit number.

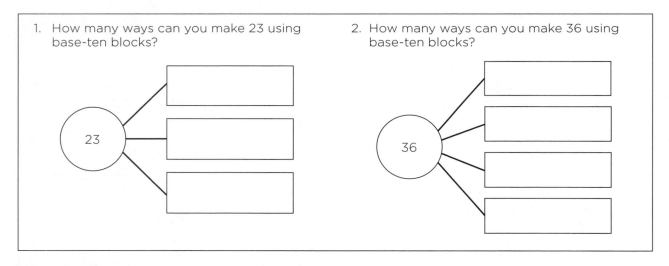

1. How many ways can you make 23 using base-ten blocks?

2. How many ways can you make 36 using base-ten blocks?

Figure 2.7: Target circle using place value.

Number Webs

Number webs are similar to target circles, but number webs allow students to make as many connections or representations as they can create to reason abstractly and quantitatively. They are similar to concept maps in literacy. In this way, number webs can serve as a diagnostic tool. Teachers can evaluate students' conceptual development by analyzing their work. Number webs can show growth. By presenting the same web at the beginning of a unit and then again at the end of the unit, students and teachers can quickly see the progress of students' learning. Figure 2.8 shows one fifth grader's work before and after a unit focusing on fractions.

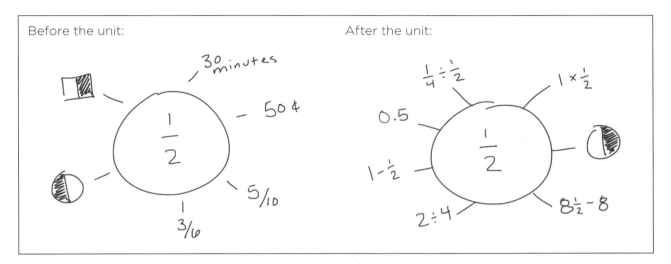

Figure 2.8: Number webs before and after a fraction unit.

In the web before the unit, the student presents more concrete representations for $\frac{1}{2}$ and also includes practical contexts such as 30 minutes and 50 cents. After the unit, the student presents more operational representations. When her teacher asks her about how she knew that $\frac{1}{4}$ divided by $\frac{1}{2}$ is $\frac{1}{2}$, the student replies, "I got this when we used the fraction strips. I see that only $\frac{1}{2}$ of $\frac{1}{2}$ fits in $\frac{1}{4}$." She is reasoning abstractly and quantitatively as she makes sense of the different representations and meanings of fractions. Teachers can use any value in the circle of the web, including mixed numbers, decimal representations, and improper fractions.

Always, Sometimes, Never

Always, *sometimes*, and *never* are choices that students select when given a specific context. Students' reasoning is featured along with the justifications for their responses. For example, you could ask grade-level students whether the following are always true, sometimes true, or never true.

- **Grade 5:** When you cut a piece off a shape, you reduce the area and perimeter of the shape (Swan, 2005).

- **Grade 4:** When you multiply a whole number by a fraction between 0 and 1, the result is greater than the starting whole number.

- **Grade 3:** When two fractions have the same numerator, they are equivalent.

- **Grade 2:** When you add two odd numbers, the sum is an even number.

- **Grade 1:** When you add three numbers all less than 20, the sum is less than 20.

- **Kindergarten:** When you add two numbers less than 10, the sum is less than 10.

Always, Sometimes, Never statements highlight students' reasoning. Students must provide the evidence that supports their claim. By providing students an opportunity to share their work with other students, both partners benefit. Students begin to see their peers as resources when they share their thinking and deepen their understanding of the mathematics. Students in such an environment do not become overly dependent on their teacher as the sole authority of information.

Just as students look for evidence in their reading, students support their mathematical reasoning by providing specific examples to justify their responses to the Always, Sometimes, Never prompts.

Always, Sometimes, Never statements can also incorporate student movement during a lesson. Place a sign in three corners of the room. One corner is for Always, one for Sometimes, and one for Never. After students have had an opportunity to think independently, they collaborate with a partner to agree on a response and justify their reasoning. When the teacher signals, students go to the spot in the room that they have selected as the correct response. The teacher randomly calls on one student in each response group, and that student explains his or her reasoning. After students have explained each answer, the teacher allows one minute for partners to determine if they still think their answer is correct or if they feel they should move. On a signal, students stay or move. To conclude this prompt, you can then conduct a whole-group discussion leading to a summary of students' thinking.

Headlines

Headlines provide opportunities for students to write in their mathematics class. Students are presented with a number sentence such as $14 = 7 \times 2$, and they are asked to create a story problem that the number sentence could solve. Students can also begin working with an expression such as 24×15 and create and solve a matching story. Figure 2.9 shows fifth-grade responses to solve 24×15.

Write a story problem that you would solve by finding the product of 24 × 15.

Student A:

Angel was riding his scooter outside. After 15 hours, he went in for lunch. The next day he rode his scooter for 24 hours. How many hours did Angel ride his scooter?

Student B:

Jo mows lawns for 15 hours.
He gets paid $24 each lawn.
How long did it take to mow
    both lawns?

Source: Adapted from Schrock, Norris, Pugalee, Seitz, & Hollingshead, 2013. Used with permission.

Figure 2.9: Students' responses to headline.

Student A demonstrates his additive thinking by asking, "How many hours did Angel ride his scooter?" He has not understood the meaning of multiplication (nor the fact that riding a scooter for twenty-four hours does not make sense). Similarly, student B sets up a context for multiplication but then demonstrates her additive thinking by asking how long it takes to mow both lawns, even though two lawns are not mentioned in the problem. Student B, however, could have demonstrated her understanding of multiplication by stating, "He gets paid $24 per hour," (not per lawn) and then asking how much money Jo earned mowing lawns.

Headlines provide an opportunity for teachers to gain insights into students' conceptual understanding. As this example illustrates, the students have not understood the meaning of multiplication even though both students can calculate 24 × 15. Having experiences that build the conceptual understanding of multiplication is critical for the development of mathematical proficiency.

Show You Know

Teachers can use Show You Know cards as a formative assessment tool. Show You Know cards provide students with limited choices such as greater than, less than, equal to, or cannot tell. Teachers write the options on the front and back of an index card for each student. The card is folded down the center so that the teacher and students see the same squares in the same position. Figure 2.10 provides an example.

Greater Than	Less Than	Less Than	Greater Than
Equal	Cannot Tell	Cannot Tell	Equal

| **Front of Card** | | **Back of Card** | |

Figure 2.10: Show You Know cards.

The teacher presents a question, and students respond by holding the card in such a way that their finger and thumb correspond to the selected answer on both sides of the card.

For example, a fourth-grade teacher might ask:

- "How does 0.25 compare to 0.10?"
- "How does $\frac{1}{2}$ compare to five-tenths?"
- "How does 14.5 compare to 145?"

Teachers can have students turn to their elbow partner and share their reasons for making their selection. Teachers can use students' responses as indicators of necessary next steps.

Think-Alouds

As mentioned in chapter 1, students can use a think-aloud to share their thinking with a partner. Similarly, think-alouds enable teachers to share their thinking as they consider a particular question. This technique models a strategy students can use when solving problems. Careful planning is a prerequisite for success.

Suppose you want your third graders to work with elapsed time. You might model a think-aloud in the following manner.

> "I'm wondering how many minutes we have until we go to music. I am going to think out loud so you can hear what I am thinking. Pay close attention to the order of my thinking so we can talk about it later. Ready?
> "I know that we leave for music at 11:10. It is now 10:35. I think I can find the number of minutes by counting by fives until I reach 11:10. Hmm. There are two groups of five until 10:45. Then, there are three groups of five from 10:45 to 11:00 and then two more groups of five from 11:00 to 11:10. So, there are seven groups of five between 10:35 and 11:10, and seven groups of five minutes are the same as 7 × 5 or 35. There are thirty-five minutes until we leave for music."

At this point, you can have students discuss with their partners the steps that you took in your think-aloud.

Number Talks

Number talks provide opportunities for students to share their strategies as they work mentally. As Sherry Parrish (2010) writes, "Mental computation is a key component of number talks because it encourages students to build on number relationships to solve problems instead of relying on memorized procedures" (p. 13).

To implement number talks, you pose a problem that students are expected to solve mentally. Students use hand signals to let you know they have a strategy and are working to find an answer. They place a fist on their chests as they think. When they think of one strategy, they show a 1 with their index finger while holding their hand to their chest. If they think of two strategies, they show two fingers, and so on. When students are ready, one student offers an answer to the problem and how he or she thought about solving it. You can record that student's thinking. After several students have shared their methods, the class can then consider the efficiency of the methods or how easy a method might be. Students can discover connections between strategies. You can save students' thinking on chart paper and refer to their strategies over the course of the unit.

The purpose of number talks is not to teach a particular strategy but to understand what students know and are able to do and for students to learn from one another. Students enjoy number talks, as they share their thinking and hear about strategies other students are using. As Jo Boaler (2016) states, "Number talks are the best pedagogical method I know for developing number sense and helping students see the flexible and conceptual nature of math" (p. 50).

Lesson Example for Mathematical Practice 2: Total Unknown or Part Unknown

The lesson plan in figure 2.11 focuses on reasoning abstractly and quantitatively. Students will solve addition and subtraction problems in which the total or part is unknown. Students first work independently to gather their own thoughts about the problem, then they share their ideas with a partner, and later they share ideas in groups of four. Throughout this lesson, teachers listen to students' discussions, noting what strategies and understandings to share with the class in the close of the lesson. This lesson is designed for second-grade classrooms but could also be used with third-grade students. A commentary follows the lesson providing more information related to the rationale and importance of each lesson component. Figures 2.12 through 2.15 (pages 58–59) support the lesson's tasks.

Unit: Solving word problems (2.OA.1)

Date: October 8

Lesson: Add to and take apart (total unknown, part unknown)

Learning objective: As a result of class today, students will be able to solve addition and subtraction problems and share their strategies.

Essential Standard for Mathematical Practice: As a result of class today, students will be able to demonstrate greater proficiency in which Standard for Mathematical Practice?

Mathematical Practice 2: "Reason abstractly and quantitatively."

- Students will read and make sense of word problems.
- Students will solve problems and explain their thinking to others.

---

continued →

Formative assessment process: How will students be expected to demonstrate mastery of the learning objective during in-class checks for understanding teacher feedback, and student action on that feedback?

- Students will use red and green cups to indicate their level of confidence as they work through the problems.
- Students will share their thinking with their partner and then again in their group of four while teachers circulate to monitor students' progress and listen to students' strategies.

Probing Questions for Differentiation on Mathematical Tasks

Assessing Questions	Advancing Questions
(Create questions to scaffold instruction for students who are stuck during the lesson or the lesson tasks.)	(Create questions to further learning for students who are ready to advance beyond the learning standard.)
• What is the problem asking? • What action is occurring in this problem? • What do the numbers represent in this problem? • What strategies might you consider using (use objects, make a picture, part-part-whole, number line, or other graphic organizer)?	• Does your answer make sense? • How can you represent this problem in a different way? • 14 + ? = 72 This is the headline. Write a story problem that you could use the equation to solve.

Tasks (Tasks can vary from lesson to lesson.)	What Will the Teacher Be Doing? (How will the teacher present and then monitor student response to the task?)	What Will Students Be Doing? (How will students be actively engaged in each part of the lesson?)
Beginning-of-Class Routines How does the warm-up activity connect to students' prior knowledge, or how is it based on analysis of homework?	Teacher will conduct a number talk. Remind students of hand signals for number of strategies. 6 + 6 = ? Ask students to share answer and strategy with their elbow partner. Teacher records one or two strategies. 16 + 16 = ? After students solve and share with partners, ask for strategies and "How did the first question help you answer the second one?"	Students sit in front of the whiteboard and mentally find the answers to the series of questions. Students share with partners. Students reflect and share with partners.

Task 1 How will students be engaged in understanding the learning objective? (See figure 2.12.)	Have students sit in groups of four. Share learning target with students. Tell them that they will work independently at first and then share with partner. When the partners agree, they will share with two other students who are in their group of four. Students share and compare their results. Remind students of the purpose of red and green cups. Green cup indicates that they are proceeding with the work. Red cup indicates that the team has a question. Actively listen to students' discussions. Look for red cups that students place on their desk indicating that both partners have a question.	Students work independently on task 1. When partners have finished, they compare their strategies and their solutions. If solutions disagree, partners work together to find the correct answer. If strategies are different, students share their reasoning. When partners are finished, they work with two other students in their group to share and compare. Students use red and green cups as indicators of progress.
Task 2 How will the task develop student sense making and reasoning? (See figure 2.13.)	For those groups of four who have completed task 1, provide them with the task 2 activity, Pick a Pair (see figure 2.13, page 58). Actively listen to students' discussions. Record statements that you want to revisit with the whole group.	Students select values they will use to place in the problem. They will then solve the problem. Students discuss in their group how the problems are the same and how the problems are different.
Task 3 How will the task require student conjectures and communication? (See figure 2.14.)	Provide partners with task 3. Students work independently at first and then share and compare with their partners.	Students work independently to determine if the problem's solution is correct. When the partners collaborate, each student needs to be ready to justify his or her thinking.
Closure How will student questions and reflections be elicited in the summary of the lesson? How will students' understanding of the learning objective be determined? (See figure 2.15.)	Ask students to come together, bring their work, and sit with their partner. Ask, "In task 1, how were these two problems the same? How were they different?" Record unique responses on chart paper. Ask, "Did you use the same operation to solve both problems?" Record class conclusions. Review learning target.	Students discuss with partners. Selected students share with class. Students discuss. Students indicate on the Likert scale their level of understanding in terms of the learning target.

Source: Template adapted from Kanold, 2012c. Used with permission.

Figure 2.11: Grade 2 lesson-planning tool for Mathematical Practice 2.

*Visit **go.solution-tree.com/MathematicsatWork** for a free reproducible version of this figure.*

Task 1: Problems for Partners

Name: _____   Partner's name: _____

Problem 1

Jeremy brought 18 baseball cards to his friend's house. His friend gave Jeremy 21 more cards. How many cards does Jeremy have now?

Problem 2

Alicia has 24 cards for her game. Her friend gave her some more cards. Alicia now has 38 cards. How many cards did Alicia's friend give to her?

Figure 2.12: Task 1 for Mathematical Practice 2 grade 2 lesson.

Visit go.solution-tree.com/MathematicsatWork for a free reproducible version of this figure.

Task 2: Pick a Pair

You are given pairs of numbers to choose from when solving each problem. Circle the pair of numbers that you would like to use. Then, write the numbers in the problem and solve it. Be sure to show your thinking.

Number Pair Choices: 8 and 17   26 and 41   59 and 82   71 and 98

1. Peter has _____ dollars in his bank. He earned more money doing work for his neighbor. When he got home, he put the money in his bank. Peter counted all of his money, and he found that he had _____ dollars. How much money did Peter earn working?

Number Pair Choices: 12 and 7   26 and 11   53 and 27   86 and 49

2. Natisha is saving to buy her mother a birthday present. The present costs _____ dollars. Natisha currently has _____ dollars. How much more money does she need?

Figure 2.13: Task 2 for Mathematical Practice 2 grade 2 lesson.

Visit go.solution-tree.com/MathematicsatWork for a free reproducible version of this figure.

Task 3: Do You Agree With Me?

I have some lollipops for my birthday party, and I put them in a bowl. My mother gave me 26 more lollipops to put in the bowl. I now have 42 lollipops! How many lollipops did I have in the beginning?

Read Kobe's work.

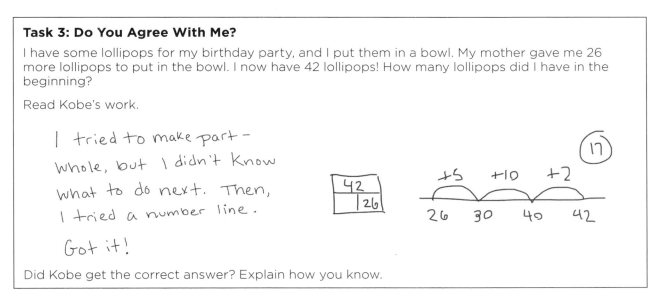

Did Kobe get the correct answer? Explain how you know.

Figure 2.14: Task 3 for Mathematical Practice 2 grade 2 lesson.

*Visit **go.solution-tree.com/MathematicsatWork** for a free reproducible version of this figure.*

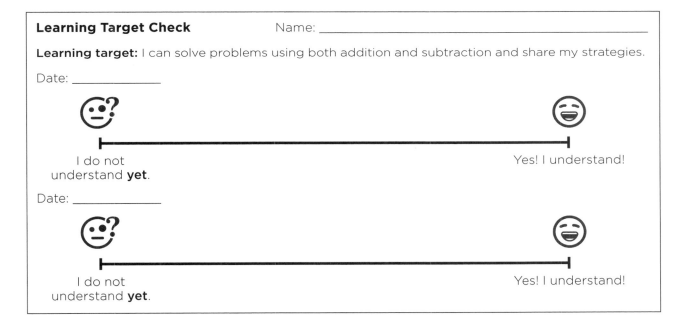

Figure 2.15: Closure task for Mathematical Practice 2 grade 2 lesson.

*Visit **go.solution-tree.com/MathematicsatWork** for a free reproducible version of this figure.*

The lesson opens with a number talk. This whole-class activity features mental computations and the sharing of strategies. The next task asks students to solve two problems. As they work, partners place red and green cups in a stack on their desks. If the green cup is on the top of the two cups, this indicates to the teacher that these students are moving forward. If a red cup is on the top of the stack, it indicates that they have a question. This task concludes by asking the partners to share their reasoning with each other.

The second task, titled Pick a Pair, offers an opportunity for students to select the size of the numbers that they would like to use to solve the problem. The size of the values of numbers selected can serve as a diagnostic tool for teachers. A student who consistently selects the larger values may have more confidence than a student who only chooses the smallest values. You may want to ask a student why he or she chose a particular pair of numbers. At the end of this task, students compare the two problems' characteristics.

Task 3 requires that students analyze a problem. At first, students need to read and analyze Kobe's solution. Students then determine whether Kobe made an error. This problem varies in that the number of lollipops at the start is unknown.

The closure for this lesson enables students to reflect on the types of problems being presented. Teachers facilitate the discussion and look for students to reason that sometimes the total is not known and other times the part is not known. Students can then discuss which operation they prefer to use in each case. Does everyone use the same operation? Why is it possible to use different operations?

The lesson closure also includes an opportunity for students to reflect on their progress toward the stated learning target presented at the beginning of the lesson. Students place an X on the scale to reflect their understanding at this point in time. Students should use this sheet repeatedly to indicate their progress toward mastery of the learning target. Discuss with students how they also reasoned about quantities during the lesson.

Summary and Action

Mathematically proficient students reason abstractly and quantitatively as they solve problems, justify their reasoning, and make sense of numbers and the given context. Today's learners are more actively involved in their learning as teachers intentionally plan to provide experiences for students to think, reason, and draw conclusions.

Identify a content standard you are having students learn currently or in the near future. Choose at least two of the Mathematical Practice 2 strategies from the following list to develop the habits of mind in students in order to reason abstractly and quantitatively.

- Quantity questions
- Graphic organizers for reasoning about quantities
    - Venn diagrams
    - Target circles
    - Number webs
- Always, Sometimes, Never
- Headlines
- Show You Know
- Think-alouds
- Number talks

Record these in the reproducible "Strategies for Mathematical Practice 2: Reason Abstractly and Quantitatively." (Visit **go.solution-tree.com/MathematicsatWork** to download this free reproducible.) How were all students engaged when using the strategy? What was the impact on student learning? How do you know?

Chapter 3

Standard for Mathematical Practice 3: Construct Viable Arguments and Critique the Reasoning of Others

Higher-order thinking thrives on interaction.

—PÉRSIDA HIMMELE & WILLIAM HIMMELE

Subitizing, the ability to quickly determine a small number of objects without having to count, enables young students to connect number names with quantities and begin to develop their sense of number, which leads to an understanding of addition and subtraction. Students practice subitizing when they quickly state the number shown using different configurations of ten frames, dots on a number cube, tallies, or base-ten blocks, to name a few. The following scenario illustrates the power of subitizing.

While working in a first-grade classroom, we used two ten frames to help students practice finding sums within 20. The frames appeared on the SMART Board as quick images. Students looked at the images (figure 3.1) and then shared with a partner how many dots they saw and how they saw them. The images remained on the screen for only two or three seconds so students would use strategies other than counting.

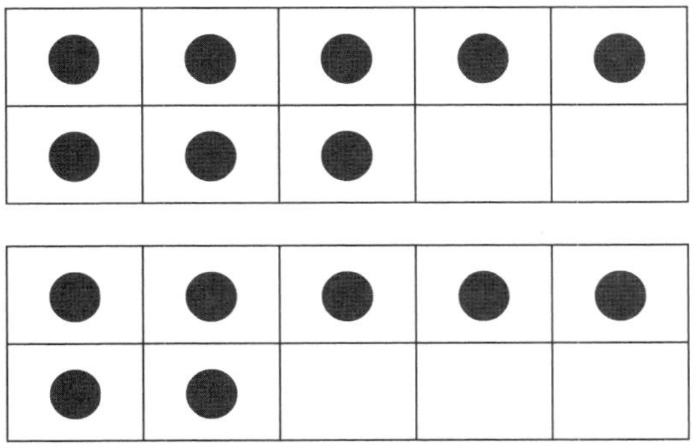

Figure 3.1: Ten frames for quick images.

After the image in figure 3.1 was projected, two students, Ralph and Enrico, engaged in a rather heated debate.

> **Ralph:** That's cheating. You can't move the dots!
>
> **Enrico:** I didn't move them. I just thought about moving two dots from the bottom frame, and I put them in the top.
>
> **Ralph:** That means that you changed the numbers. You can't do that.
>
> **Enrico:** I didn't put more chips on so the total is the same. The frames look different in my head than when I first saw them.

At this point Kit asked the boys if they would be willing to share their thinking with the rest of the class. She showed the original two ten frames and asked Enrico to share his strategy and show the class how he thought about finding the sum of 8 + 7. Enrico explained that adding a number to 10 was easy, so he made a 10 by using 2 dots from the 7. Enrico moved the dots on the board to show a filled frame on the top with 5 remaining in the lower frame. Ralph then said excitedly, "That means that 8 + 7 is the same as 10 + 5. Wow, that is so easy!"

In this case, Enrico made a convincing argument, and Ralph and the other students added his strategy to those they already think about when finding sums. What we originally perceived as a social argument between two students became an opportunity to create and share a quality mathematical argument.

Mathematics teachers can no longer be satisfied that students simply arrive at correct answers. It is important to ensure that students not only produce correct answers but do so for the right reasons. Consider the following scenario.

Carolyn was asked to write 356,279 in expanded form. She says, "I like these types. I just write the numbers separated from each other, and then I place more zeros each time after the first one." Carolyn writes the digits starting on the right, as indicated by the arrow. She places the next digit and one zero to the left of the first value and continues to the next value, placing one more zero each time (see figure 3.2).

Figure 3.2: Carolyn's expanded form example.

Carolyn is not thinking about place value as she writes numbers in their expanded form. She is getting the correct answers, but for the wrong reasons. Her thinking is procedural, and she would not likely be able to construct a convincing argument as to why her method works. She might justify her approach by saying, "I don't know why this works; it just does."

The high-level task in figure 3.3 asks students to critique another student's approach to adding two fractions.

Grade 5

Same or Different?

Fifth-grade students at Pascal's Elementary School are working on fractions. The teacher began class by asking students to solve this problem.

$$\frac{2}{3} + \frac{1}{4}$$

Zoe said, "I am thinking that 12 represents the whole. So, I can find $\frac{2}{3}$ of 12, and then I can find $\frac{1}{4}$ of 12. Next, I just add the two quantities.

$\frac{2}{3}$ of 12 is the same as 8, and $\frac{1}{4}$ of 12 is the same as 3. I know that 8 + 3 =11.

$$\frac{2}{3} + \frac{1}{4} = \left(\frac{11}{12}\right)$$

You are Zoe's partner, and she shared her work with you. She is not sure her strategy works.

Solve the problem a different way.

$$\frac{2}{3} + \frac{1}{4}$$

What do you think of Zoe's strategy? Does it make sense? Be specific as you explain your thinking.

Figure 3.3: Same or Different?

*Visit **go.solution-tree.com/MathematicsatWork** for a free reproducible version of this figure.*

As the sample student response to this prompt in figure 3.4 illustrates, the student's thinking is procedural. She is using what some teachers refer to as the *butterfly method*. She multiplies the denominator, 3, times the numerator in the other fraction, $\frac{1}{4}$. She then multiplies the denominator of the first fraction by the numerator of the second fraction. She adds those two products, 3 and 8, and then multiplies the denominators, though the lack of precision in the work makes the thinking hard to decipher. This gives her the same answer as Zoe, $\frac{11}{12}$. This student's rationale is simply that she got the same answer, so Zoe must be correct. Students struggle to explain this butterfly method to add fractions. This method simply does not make sense to most students.

Solve the problem a different way.

$\frac{2}{3} + \frac{1}{4}$ $3 + 8$

$\frac{1}{4} + \frac{2}{3}$

$\times 3 = \frac{11}{12}$

What do you think of Zoe's strategy? Does it make sense? Be specific as you explain your thinking.

Her strategy is correct because I got the same answer.

Figure 3.4: Same or Different?—student work.

The third Standard for Mathematical Practice, "Construct viable arguments and critique the reasoning of others," focuses on students communicating their thinking in a logical manner, both orally and in writing. In an environment that encourages taking risks, students can learn to justify their thinking, share their ideas with others, and consider the ideas of their classmates. In order to communicate effectively, mathematically proficient students use appropriate vocabulary. They look for similarities and differences among various approaches to the same problem and draw conclusions as to the efficiency and accuracy of each approach. As students mature, they can justify their reasoning in writing as well as present their arguments orally. As evident in the previous chapters, the Standards for Mathematical Practice are interwoven. Mathematical Practice 3 depends on students working with meaningful tasks (Mathematical Practice 1) that promote the need to reason abstractly and quantitatively (Mathematical Practice 2).

Table 3.1 shows more examples for what students do during a lesson to demonstrate evidence of learning Mathematical Practice 3 and what actions you can take to develop this critical thinking in students.

Table 3.1: Student Evidence and Teacher Actions for Mathematical Practice 3

	Student Evidence of Learning the Practice	Teacher Actions to Engage Students
Construct viable arguments and critique the reasoning of others.	Students: • Use examples and nonexamples to present their argument • Use diagrams, models, drawings, objects, charts, and graphs • Use appropriate vocabulary • Justify their thinking in words or calculations • Ask other students questions such as "Why?," "What do you mean?," and "Can you explain that in another way?" as they look to clarify their own understanding • Restate what they heard from others using their own words • Listen to and reflect on other students' arguments	Teachers: • Create a safe learning environment that promotes risk taking and learning from mistakes • Provide tasks that require high levels of cognitive demand • Give tasks worthy of discussions and that authentically lead to meaningful conversations • Structure opportunities for student discourse • Prepare questions in advance that promote students' thinking • Actively listen for students' understanding • Select and sequence students to present their work to the class • Explicitly make connections from one idea to the next • Model how to make an argument

*Visit **go.solution-tree.com/MathematicsatWork** for a free reproducible version of this table.*

Understand *Why*

Mathematical arguments are logical, clear statements using numbers, pictures, and words to support a given solution or conclusion. They go beyond a typical student's statement such as "I just know it" or

"This works." A mathematical argument develops as a series of steps with each subsequent step following naturally from the first. We had the privilege of eavesdropping as one second grader shared his thoughts about patterns when adding two odd numbers. He offered the following argument to his partner.

> I know that 1 is odd and 1 + 1 is even because I have a pair. I can show that with chips.

> Look what happens if I start with 3. I can arrange 3 dots this way.

> One dot does not have a match. That means 3 is odd.
>
> When I put three more dots, each one has a partner! 3 + 3 is even. Two odds make evens!

In years past, students' experience with proofs began in high school geometry. Now, elementary educators have featured the importance of logical reasoning and justification of solutions. According to Andreas Stylianides (2007):

> The most important reason for this growing emphasis is that proof and proving are fundamental to doing and knowing mathematics; they are the basis of mathematical understanding and essential in developing, establishing, and communicating mathematical knowledge (e.g., Hanna & Janhnke, 1996; Kitcher, 1984; Pólya, 1981). (p. 289)

It is also clear that students cannot justify and explain what they do not understand. As Susan O'Connell and John SanGiovanni (2013) express, "Understanding, or lack of understanding, is revealed when students are asked to defend their thinking" (p. 45). In this way, teachers can use students' arguments as diagnostic tools. By reflecting on the strengths and weaknesses in students' arguments, teachers can plan next steps to move students' learning forward.

 In order for students to develop the skill of creating mathematical arguments, teachers need to establish an environment that is conducive to taking risks and be certain that students see mistakes as tools for further learning. In literacy, mistakes are frequently referred to as "miscues." Maggie Siena (2009) notes that miscues are "cues to student thinking" (p. 66). In this way, teachers can use miscues as diagnostic tools. For students to feel that they can take risks, the classroom climate must be receptive and view errors as opportunities.

In *The Skillful Teacher*, Jon Saphier and Robert Gower (1997) offer the following list that characterizes five areas of the classroom climate that contribute to students' ability to learn, followed by a student's explanation of the item.

1. **Knowing others:** "I know these people and they know me."

2. **Greeting, acknowledging, listening, responding, and affirming:** "I feel accepted and included. People respect me, and I respect them."

3. **Group identity, responsibility, and interdependence:** "I'm a member of this group. We need each other and want each other to succeed."

4. **Cooperative learning, social skills, group meetings, and group dynamics:** "I can help others and they will help me."

5. **Problem solving and conflict resolution:** "We can solve problems that arise between us." (p. 361)

These relationships and sense of belonging do not develop overnight. Teachers must actively pursue these qualities. Teachers who take the time and energy to invest in these areas of classroom climate realize the benefits in terms of students' learning. Students who feel safe enough to take risks, share their thinking with others, and feel that they belong in the class open themselves to new ideas and explorations. (For more details on building classroom climate, see chapter 13 in *The Skillful Teacher* [Saphier & Gower, 1997].)

Research suggests that teachers should limit the quantity of classroom rules. According to Carolyn Evertson, Edmund Emmer, and Murray Worsham (2003), elementary teachers should develop five to eight rules and expectations. Two examples of norms and agreements are as follows.

Group work norms:

*   Everyone has a right to ask for help.

*   When asked for help, each group member responds by offering help but not answers.

*   No one is done until everyone is done.

*   Everyone helps.

*   Everyone cleans up.

Class norms:

*   Be prepared.

*   Be respectful.

*   Be safe.

Referring to the norms and identifying when students are demonstrating the norms help students meet the expectations. You can feature a specific norm at the beginning of the lesson and then identify students' actions meeting that norm at the close of the lesson. Students can create a list of specific examples of when a norm is being followed as well as nonexamples when a norm is not implemented. Such strategies enable students to understand that the norms are important and not just a poster on the wall.

The establishment of norms allows for students to engage in the focus of this chapter, construct viable arguments and critique the reasoning of others. As noted previously, in *Principles to Actions* (NCTM, 2014), the authors identify eight Mathematics Teaching Practices. Two of these practices connect directly with this chapter's emphasis: "Facilitate meaningful mathematical discourse" and "Elicit and use evidence of student thinking" (NCTM, 2014, p. 10).

As teachers collaborate to plan units and create lesson plans, keeping track of what students are doing and documenting what teachers are doing provide a framework for reflection. Are students thinking, discussing, sharing their strategies, and presenting their arguments? Are teachers demonstrating strategies as students watch? Who is doing the work throughout the lesson? Teachers do the heavy lifting as they prepare before class; students should do the heavy lifting in class.

In order to meet the challenge of success for all students, teachers need to approach their class time differently. Teachers should give students time to engage in dialogue and share their thinking, monitor these discussions by listening and observing, and then strategically select students to explain their understandings to the entire class. Helping students make connections among those strategies that are shared and continuing to ask questions will push students' thinking. By supporting students to create mathematical arguments, the source for whether a solution is correct transfers to sound student reasoning rather than the teacher's voice. Magdalene Lampert and Paul Cobb (2003) present the value of students' discussions:

> Discourse should not be a goal in itself but rather should be focused on making sense of mathematical ideas and using them effectively in modeling and solving problems. The value of mathematical discussions is determined by whether students are learning as they participate in them. (p. 194)

A classroom climate that encourages students listening to each other and sharing their own thinking is one in which the teacher avoids saying anything that the student might say. If you rephrase what a student has just said, the message received by the other students is that they do not have to listen to the student's statement. Here are a few suggestions Max Ray (2013) offers for teachers to encourage active listening by students:

- Make sure multiple students respond to each question before asking another.
- Ask questions that have multiple possible answers, such as "What do you think is the most useful way to represent this, and why might someone else represent it differently?
- Encourage students to reflect back what they are hearing or offer alternate ways to say the same thing, as well as asking, "Is this another way to say what you are saying?" or "Am I understanding you correctly to mean this?"
- Listen *to* students' ideas, don't listen *for* correct thinking. (p. 27)

It is interesting to note that in *Principles to Actions* (NCTM, 2014), all eight Mathematics Teaching Practices focus on action steps teachers implement so students can develop the habits of mind from the Standards for Mathematical Practice, and in this case, Mathematical Practice 3: "Construct viable arguments and critique the reasoning of others."

Strategies for *How*

Creating mathematical arguments may be new to students and teachers alike. For decades, arriving at the correct answer was sufficient. Unfortunately, students can state the correct answer for the wrong reasons or not understand *why* an answer is or is not correct. In order to achieve the skills necessary for the 21st century, students need to go beyond the right answer. They need to justify their reasoning and provide a logical argument supporting their work.

Consequently, teachers must devote time in the classroom for students to share, discuss, question, draw conclusions, and justify their thinking. Consider the guideline that 65 percent of the time, students should be engaged in doing mathematics through peer-to-peer discourse (Kanold, 2015c). In a ninety-minute block of time, students should be discussing, explaining, verifying, and justifying their thinking for approximately sixty minutes. Similarly, in a sixty-minute mathematics block, students should be engaging in the Standards for Mathematical Practice for forty minutes. This does not mean that these minutes occur consecutively. (Refer to the lessons at the end of the chapters for ways to deliver this 65 percent guideline.) By providing students with this time to discuss and share their thinking, students build their understanding and have opportunities to deepen their knowledge as they construct viable arguments. The strategies that follow show students how to construct viable arguments and critique the reasoning of others.

Answers First

Answers First provides a format in which students justify the solution that is provided for them. By working with the correct answer at the start of the task, students determine why the answer is correct and how they know that to be the case. Teachers can give the answer on the task itself or post the answers for the classwork on the board. This turns the table for students. The answer is no longer sufficient; the emphasis is now placed on the work leading to that answer. Some example tasks for grades 1–5 that lead students to this thinking are shown in table 3.2.

What Doesn't Belong?

Another strategy that encourages students to justify their reasoning by creating a mathematical argument is What Doesn't Belong? Students are presented with four values or expressions, and they determine how three out of the four are connected. Students then present an argument as to why the fourth value does not belong. Consider the following example in figure 3.5.

Table 3.2: Answers First Example Tasks

Grade	Answers First Task
1	I saw lots of apples in the basket this morning. Then, 8 students each took one apple during snack. Now, I see 12 apples in the basket. How many apples were in the basket this morning? Answer: 20 apples Explain why this answer is correct. Use pictures, numbers, and words.
2	I have 3 nickels, 6 dimes, and 2 pennies in my pocket. How much are these coins worth altogether? Answer: 77 cents Why is this answer correct? Explain or show how you know.
4	Explain why $\frac{3}{10} + \frac{3}{100} = \frac{33}{100}$.
5	Compare using <, >, or =. 0.20 ◯ 0.200 Correct Answer: ⊜ Why is this answer correct? Explain or show how you know.

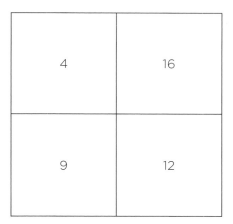

Figure 3.5: What Doesn't Belong? example.

In this case, students might say that the 9 does not belong as this is the only odd number. Students might also argue that the 9 does not belong as all of the other values are multiples of 4. Other students might say that the 12 does not belong as it is not a perfect square.

Figure 3.6 (page 72) offers another example.

In this case, the student might state, "Difference does not belong because finding the difference makes values smaller. The other words are operations that produce larger values when whole numbers are used." Another student might select *multiply* and explain that multiply is an operation to do while the other words are final answers after doing an operation.

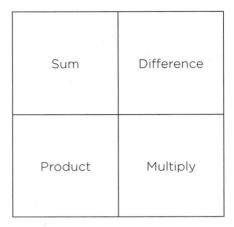

Sum	Difference
Product	Multiply

Figure 3.6: What Doesn't Belong? example with operations.

What Doesn't Belong? challenges students to create mathematical arguments for their selection and offers opportunities for students to enhance their listening skills and attend to precision as they make sense of arguments presented by their peers.

Sentence Frames

As students begin to collaborate, they may not know how to share their thoughts. Consider creating an anchor chart and posting it in the classroom for students to refer to as they work with partners and participate in classroom conversations. Some suggestions for sentence frames include the following.

- I know because _____.

- Can you explain _____?

- I agree with you because _____.

- I disagree with you because _____.

- I understand this part, but I am confused by _____.

- Thank you. You helped me to _____.

- This _____ (answer, step, or part) does not make sense because _____.

- I still have a question about _____.

- I notice _____.

- I think that there is another way to _____.

- That leads me to think about _____.

- I wonder if _____.

As students continue to collaborate, they can make suggestions to the sentence frame list and add those to the chart.

Frames can also include several blanks that students will complete based on given vocabulary or the context presented, as the next few activities illustrate. The following is a kindergarten or grade 1 example.

> I know that _____. This means that _____.
> I know because _____.

What's Under the Cup?

What's Under the Cup? is a kindergarten or first-grade activity that uses the sentence frame strategy. Partners gather a small number of chips and a cup. Both partners know the number of chips. For this example, suppose that students are working with seven chips. One student hides some chips under the cup. The other student looks at the remaining chips outside the cup. This student then uses the following sentence frame to help state what is under the cup, saying, "I know that there are 3 chips outside the cup. That means that there are 4 chips under the cup. I know because $3 + 4 = 7$ ($7 - 4 = 3$ or $7 - 3 = 4$)."

Reach 100

Students in grade 2 use sentence frames to record their results as they play Reach 100. Students use a deck of cards that show various amounts in terms of base-ten blocks. Each student selects two cards and then uses the sentence frames in figure 3.7 to record his or her work.

My first card shows _____. The card has _____ tens and _____ ones.
My second card shows _____. The card has _____ tens and _____ ones.
The sum of my addends is _____.
I found the sum by _____.
My partner's sum is _____. My sum is _____.
_____ is closer to 100. We know because _____.

Figure 3.7: Reach 100 sentence frames.

*Visit **go.solution-tree.com/MathematicsatWork** for a free reproducible version of this figure.*

If the number remains small after selecting two cards, a student may pick from the deck of cards again. The sentence frame needs slight adjustment in this case (see figure 3.8).

My first card shows _____. The card has _____ tens and _____ ones.
My second card shows _____. The card has _____ tens and _____ ones.
My third card shows _____. The card has _____ tens and _____ ones.
The sum of all three addends is _____.

Figure 3.8: Reach 100 adjusted sentence frames.

*Visit **go.solution-tree.com/MathematicsatWork** for a free reproducible version of this figure.*

I Know This, So I Know That

Sentence frames can also help students recognize patterns. I Know This, So I Know That helps students see patterns between numbers. Figure 3.9 shows an example focusing on multiplying by multiples of 10 in grade 3.

Example: I know 2 × 3 = 6, so I also know that 20 × 30 = 600.

Your turn. Fill in the blanks using the same pattern as in the example.

I know that __3__ × __3__ = _____ so I also know that _____ × _____ = _____.

I know that __4__ × __2__ = _____ so I also know that _____ × _____ = _____.

Now select the numbers in the first blanks and follow the pattern to fill in the rest of the blanks:

I know that _____ × _____ = _____ so I also know that _____ × _____ = _____.

I know that _____ × _____ = _____ so I also know that _____ × _____ = _____.

Describe the pattern in these examples. Explain why your answers are correct.

Share your work with a partner. Are your explanations both correct? Why or why not?

If you need to make additions or changes to your explanation, do so here.

Figure 3.9: I Know This, So I Know That sentence frames.

*Visit **go.solution-tree.com/MathematicsatWork** for a free reproducible version of this figure.*

Written Explanations

Another way to engage students in constructing arguments and critiquing the reasoning of others occurs through written explanations. When students can articulate their thoughts or the ideas of others on paper, the process deepens their understanding. As Marilyn Burns (1995) states, "Writing encourages students to examine their ideas and reflect on what they have learned. It helps them deepen and extend their understanding. When students write about mathematics, they are actively involved in thinking and learning about mathematics" (p. 13).

Written explanations do not have to be lengthy, seemingly endless paragraphs. When standardized state tests were first implemented, some instructors felt that students needed to not only rephrase the question but also write out every step using words *and* numbers. Students tended to think that there was a direct correspondence between the number of words they used and their score. However, it was never about the quantity of words. Rather, the emphasis should always have been and continues to be on students showing clear and concise steps in a cohesive argument that contains symbolic representations as well as verbal statements of support.

To illustrate this point, consider this third-grade task and sample student response.

Task: Latoya is making pies. She uses 5 apples and 4 pears in each pie. If she wants to make 3 pies, how many apples and pears will she need?

Response: I know she is making pies. I need to find how many apples and pears she needs.

I can add 3 groups of 5 apples; 5 + 5 + 5 = 15.
I can add 3 groups of 4 pears; 4 + 4 + 4 = 12.
She needs 15 apples and 12 pears to make the pies.
I can also use multiplication to find the answer: 3 × 5 = 15 and 3 × 4 = 12.
I get 27 pieces of fruit each time.

The written explanation makes students' thinking visible, which provides opportunities for teachers to gain insights into students' reasoning. This allows written explanations to be a formative tool.

Problems Without Numbers (adapted from Gillan, 1909) is an approach to help students focus on the process of solving problems and constructing a convincing argument. Students are not given values to substitute into algorithms. Instead, they need to make sense of the context and then formulate a solution pathway. Their written explanation serves as a window into their thinking, as in the following grade 4 example.

Task: You know how much one-half of John's money is and how much one-fourth of Ned's money is. How can you find out how much they both have together?

Sherri's response: I used play money. I made up numbers for John and Ned. John had $10, and Ned had $4. I put the money in two piles. Next, I doubled John's pile. I also doubled Ned's pile and then doubled that money again. Finally, I pushed the piles together and counted.

From Sherri's response, her teacher knows she is a concrete thinker who relies on counting all to determine the total. Her thinking and understanding about the relationship between the unit fractions and each boy's total amount of money are strong. Sherri understood that because she knows one-half of John's total, she could double that amount to find how much John had altogether. Similarly, she doubled

Ned's money twice to get his total, effectively multiplying his amount by four. A second student response is as follows.

> **Anastasia's response:** I know ½ of John's money. I can double that amount to get his total.
> For Ned, I will multiply his money by 4 to get the total.
> Now I add these two together to find all of the money.

Anastasia's thinking is similar to Sherri's. They both understand the fractional part of the whole. Anastasia is now thinking more symbolically and does not need to represent the amounts of money using concrete objects or amounts. Both students' arguments are logical and will lead them to correct answers. These students would benefit from an opportunity to compare and contrast their explanations and to critique each other's arguments.

Work Comparisons

Another strategy to engage students in constructing viable arguments and critiquing the reasoning of others involves comparing work. Providing opportunities for students to analyze similarities and differences in work samples enriches their understanding of the concepts. They learn by comparing their own work and understanding to that of a teacher or another student. Three examples of work comparisons are Find the Error, Make Suggestions, and Factor Facts.

Find the Error

One way to engage students in critiquing the arguments of others involves showing students examples of work containing mistakes. An example of comparisons from a kindergarten class follows in which students compare what they hear their teacher say and what they know to be correct.

Teacher: I am going to count. Listen carefully. I may make a mistake. If you hear a mistake, use our quiet hand signal.*

*Note: Quiet hand signal begins with students placing their fist below their chin. When they hear a mistake, they extend one finger. If they hear a second mistake, they extend another finger.

Teacher: Ready?

1, 2, 4, 5, 7

Teacher: Did you hear a mistake? Turn and talk with your elbow partner. What mistake did I make?

Davis: We think you forgot 3.

Teacher: When should I have said "3"?

Davis: 3 comes after 2.

Lolly: 3 comes before 4.

Sarah: We found another mistake too. You missed 6, and 6 comes after 5. You skipped it.

Teacher: Thumbs up if you agree. Here is another one. Be ready, and listen carefully.

7, 8, 9, 9, 10, 11

Teacher: Turn and talk with your partner. Was there a mistake?

Teacher: Tell me what your partner thought. . . . Austin?

Austin: Millie thought you said "9" two times. You can't do that.

Teacher: Why?

Millie: Each number is said just once.

This teacher is guiding her students to begin to use logical arguments to justify their thinking. It is also reasonable to ask other students whether they agree with a statement just made by another student. It is not only the teacher who validates students' answers and thinking! In this way, students can see each other as resources and the classroom community is strengthened.

Even young students can explain their thinking and use what makes sense to them to share their strategies. All three of these kindergarten students expressed their thinking using a different approach.

When using Find the Error, it is important to not only ask for identification of the error, but an explanation for why an error was made or how to avoid it in future problems. Asking students to find errors and to write suggestions for ways to avoid the error or why the solution is incorrect deepens their understanding. The third-grade task in figure 3.10 shows an example of students using this extension to Find the Error.

Salena wrote fractions above each of the diagrams. She says the fractions name the shaded part. Do you agree with Salena? Why or why not? Provide a specific reason for your answer.

$$\frac{1}{5}$$　　　　　　　$$\frac{1}{4}$$

Figure 3.10: Find the Error activity.

*Visit **go.solution-tree.com/MathematicsatWork** for a free reproducible version of this figure.*

See Mathematical Practice 6, "Attend to Precision" (page 167) where this idea is explored further as Error Analysis.

Make Suggestions

Students can also analyze student work and compare it to a more efficient or effective strategy and then make suggestions for improvement to a solution. The following example in figure 3.11 asks grade 2 students to critique a student's approach, make suggestions as to ways to be more efficient, and then solve the problem using a different method.

Grade 2: Make Suggestions

Marshall has 22 baseball cards. His mother gave him some more for his birthday. He now has 41 baseball cards. How many cards did his mother give him?

Here is how Rashawn solved this problem:

I started with 22. I made jumps on the number line until I stopped at 41.

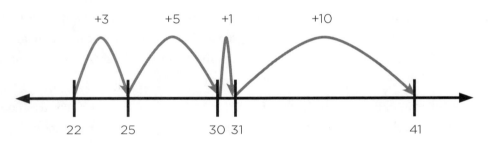

Is Rashawn's work correct? _____ Explain how you know.

Did she state her answer? _____ What is her answer?

Rashawn shows many steps. Make a suggestion to her as to how she can present her work with fewer steps.

Determine the number of baseball cards that Marshall's mother gave him using a different method.

Figure 3.11: Grade 2 Make Suggestions example.

Visit **go.solution-tree.com/MathematicsatWork** *for a free reproducible version of this figure.*

Factor Facts

A third work comparison activity is Factor Facts. Students work independently at first to determine the size of the products given the size of the factors. Students then compare their work and share their thinking. If their conclusions change, the partners take a new comparison sheet and rewrite their conclusions. Teams of two can then compare their justifications with another team. If the answers vary, each team should make a case for its own conclusions. The teams present their arguments until they reach consensus. Teachers should listen carefully to students' reasons and offer opportunities to share their thoughts with the whole class.

For example, if both factors are greater than one, then the product will be greater than both of the factors. If both factors are less than one, the product will be less than either factor. Students should decide on the size of the product and provide specific examples supporting their reasoning. Figure 3.12 shows an example of a grade 5 Factor Facts activity.

Factor Facts		
Determine the *size of the product* relative to the size of its factors if all factors are greater than 0 and . . .		
Both factors are greater than 1.	One of the factors is less than 1.	Both factors are less than 1.
Justify your answer. Offer specific examples in your argument.	Justify your answer. Offer specific examples in your argument.	Justify your answer. Offer specific examples in your argument.
The product will be greater than 1 because if both factors are greater than 1, then the answer has to be greater than 1. ex: $2 \times 2 = 4$ greater greater greater than 1 than 1 than 1	The product will be less or greater than 1. ex: $0.5 \times 1.5 = 0.75$ factor less than 1 — product less than 1 2nd ex: $0.5 \times 3 = 1.5$ factor less than 1 — Product more than 1	The product will be less than 1 because if both factors are less than 1, then the product has to be less than 1. ex: $0.5 \times 0.5 = 0.25$ less less less than 1 than 1 than 1

Figure 3.12: Grade 5 Factor Facts.

Visit **go.solution-tree.com/MathematicsatWork** *for a free reproducible version of this figure.*

In this example, Olivia has presented three clear explanations for each case, and she included an example supporting her arguments.

Peer Review

Peer review provides opportunities for students to reflect on their partner's work while deepening their own understanding. Students need to voice their understandings, hear additional and perhaps contrasting ideas to their own, and extend their thinking. Structure peer review so each partner has an equal opportunity to contribute. Peer review provides the class with a necessary structure and involves several steps, as the following nine illustrate.

1. Teacher provides time for all students to reflect on their own work. What do they want to share? What was their strategy? Does their answer make sense? Why does it make sense?

2. Partners exchange work. Each student reflects on these same questions about his or her partner's work without speaking.

3. On the teacher's signal, partner A speaks for a full minute about his or her own work. Partner B listens without saying a word.

4. After the minute, partner B shares what partner A said thus far and then poses any clarifying questions or suggestions for thirty seconds.

5. Partner A answers any questions and then summarizes what partner B said.

6. Both partners take a fifteen-second break.

7. Partner B then speaks for one minute about his or her own work by sharing the approach used and why the answer makes sense.

8. Partner A listens and then asks clarifying questions when partner B's minute ends.

9. Partner B responds to any questions and then summarizes what partner A said.

Table 3.3 shows an alternate version of peer review protocol and clarifies students' work during the process.

Table 3.3: Peer Review Protocol

Partner A	Partner B	Time
Partner A addresses the strategy he or she used, the solution, and why the solution makes sense.	Partner B listens.	One minute
Partner A listens.	Partner B shares what partner A said and then asks any clarifying questions and makes suggestions.	Thirty seconds
Partner A responds to questions and then summarizes what partner B said.	Partner B listens.	Thirty seconds
Break		Fifteen seconds

Partner A	Partner B	Time
Partner A listens carefully.	Partner B addresses the strategy he or she used, the solution, and why the solution makes sense.	One minute
Partner A shares what partner B said and then asks any clarifying questions and makes suggestions.	Partner B listens carefully.	Thirty seconds
Partner A listens carefully.	Partner B responds to questions and then summarizes what partner A said.	Thirty seconds

Visit **go.solution-tree.com/MathematicsatWork** *for a free reproducible version of this table.*

Students frequently have difficulty speaking as well as listening for the entire minute. With practice, students find that their statements offer more details about their thinking and the strategies they used to solve the problem. When asked to reflect on the use of the protocol, students typically respond by saying that the experience was frustrating at first, but then they came to appreciate hearing what their partners said about their work. Students usually find it easy to listen and hear suggestions. It is helpful to post the protocol's steps in the classroom and use a timer as students participate in peer review.

 ## Rubrics

Rubrics can serve as a tool for students' self-reflection and for use in peer reviews. Rubrics can also establish expectations for a product or writing sample. Table 3.4 (page 82) presents an example students can use to self-reflect after solving a task.

You and your students can use the rubric to identify and clarify how closely you are both evaluating the same work. Students can use yellow highlighters to indicate bullets that they think they've exhibited in their work. You can use a blue highlighter to reflect your thoughts about the same work. As yellow and blue combine to make green, all of the final highlights in green indicate agreement.

Students can also use the rubric to sort samples of student work as a class, using the classifications of strong and weak. Samples presented to the class can be created by you or your colleagues, student work from previous years or from other classes, or work from students in the current class. The decision to present current students' work depends on the classroom culture. How receptive are students to feedback from peers? Are mistakes a natural part of the learning process? If the work is weak, they discuss how to make it strong. Students can then apply learning from the discussion to improve their own work. They can use colored pencils to make changes so they look back and see how their changes improved their work. Such discussions further clarify what a viable argument is and will help students use a rubric to critique the reasoning of others.

Table 3.4: Problem-Solving Rubric

Needs More Work 1	Making Progress 2	Almost There 3	Arrived! 4
My answer is not complete.	My work shows that I have some understanding of the problem, but I left out some important work or partial answers.	My answer shows that I understand the problem adequately.	My answer shows that I understand the problem thoroughly.
My answer is incorrect.	My answer is incorrect, or my answer is correct but I have not shown any supporting evidence leading to my answer.	My answer may be incorrect due to a calculation error.	My answer is correct, and I have justified it by my thinking and work.
I have shown some calculations and may have included a chart or diagram, but these do not relate to the problem.	I have included a chart, table, or graph, but these have mistakes or are not complete.	I have included my calculations and have used a chart, table, graph, or appropriate diagram.	I included calculations as well as charts, graphs, diagrams, or drawings that illustrate the mathematics.
I have written a description, but it is not complete.	My written description has gaps and is missing important vocabulary words.	My written description is clear and uses appropriate vocabulary.	My written description is clear and concise, and I use appropriate vocabulary.
A reader may not be able to follow my work.	A reader may have trouble following my thinking.	A reader can follow my thinking even though there may be some small gaps in reasoning shown.	A reader can follow my thinking.

*Visit **go.solution-tree.com/MathematicsatWork** for a free reproducible version of this table.*

Additionally, this rubric can facilitate whole-class discussion. After the class has completed a task, you can show student work samples and ask students to determine the score. If a sample receives anything below an Arrived! (level 4), consider asking students to determine ways to improve the work. You can then return the papers to students and ask them to reflect on their own work and improve their responses based on the discussion. These discussions enable students to reflect on the characteristics of quality work and then incorporate their understandings to improve their own solutions.

After a student uses the problem-solving rubric, he or she shares the rubric with a peer reviewer. Together, they discuss strategies for revising the first student's work. The first student incorporates the suggested changes and any other necessary revisions. Together students review the work and collaborate to determine the score after making revisions. Figure 3.13 provides a reflection rubric to guide the discussion.

Name: _____		Date: _____
Title of task:		

Initial Score	Action Steps to Revise the Work	Score After Revisions

For next time, I need to remember to . . .

Figure 3.13: Reflection rubric.

*Visit **go.solution-tree.com/MathematicsatWork** for a free reproducible version of this figure.*

Once students have used a rubric a few times, consider holding a class discussion on the effectiveness of the problem-solving rubric and the reflection rubric. Are there any changes that students want to make? Are there questions you should add to help focus students' work?

Rubrics can also serve as feedback to students after an assessment or as a chart to inform their preparation for a test. Figure 3.14 shows a chart based on a grade 3 unit assessment. Teachers would give this chart to students when returning their tests that indicate correct and incorrect responses and the points the student earned for each problem.

Grade 3 Unit Assessment: Multiplication and Division				
Learning Target	**Standard**	**Test Questions**	**Total Points I Earned**	**Reflection (Circle one.)**
I can interpret products in multiplication.	3.OA.1	1, 2, 3	____ out of 6	Not Yet Yes I can!
I can interpret quotients in division.	3.OA.2	4, 5, 6	____ out of 6	Not Yet Yes I can!
I can write a problem that will be solved using multiplication.	3.OA.1	7	____ out of 3	Not Yet Yes I can!
I can write a problem that will be solved using division.	3.OA.2	8	____ out of 3	Not Yet Yes I can!

Figure 3.14: Feedback rubric.

*Visit **go.solution-tree.com/MathematicsatWork** for a free reproducible version of this figure.*

Students then review their tests and record the total number of points earned. Students also use the reflection rubric to determine where they are currently in terms of progress toward mastery of the learning target, and teachers can provide them with next steps to make progress. Using a feedback rubric formatively prevents the learning from stopping at the end of the assessment and provides students with ongoing opportunities to enhance their understanding.

Student Share

The tables have turned. Teachers are listening more and talking less throughout daily lessons. As Steven Reinhart (2000) states, "My definition of a good teacher has since changed from 'one who explains things so well that students understand' to 'one who gets students to explain things so well that they can be understood'" (p. 54). Classroom climate determines whether students feel comfortable offering their thoughts. As mentioned earlier in this chapter, the appropriate classroom climate accepts mistakes, or miscues, as learning opportunities and supports students as they offer their mathematical understanding. The appropriate climate transfers the sole responsibility for learning from the teacher to students. As mentioned in the Understand *Why* section of this chapter (page 66), Ray's (2013) sample list of suggestions to encourage active listening promotes greater opportunities for students to participate in discussions.

In order to move students' discussions toward intentional outcomes, keep the lesson objectives in mind. In *5 Practices for Orchestrating Productive Mathematics Discussions*, Margaret Smith and Mary Kay Stein (2011) define five action steps for teachers to incorporate in order to promote effective discussions.

1. ***anticipating*** likely student responses to challenging mathematical tasks;

2. ***monitoring*** students' actual responses to tasks (while students work on the tasks in pairs or small groups);

3. ***selecting*** particular students to present their mathematical work during the whole-class discussion;

4. ***sequencing*** the student responses that will be displayed in a specific order; and

5. ***connecting*** different students' responses and connecting the responses to key mathematical ideas. (p. 8)

After selecting the task for the lesson, you also need to anticipate the possible approaches that students might take. Identifying the possible misconceptions that might arise adds to teachers' readiness to implement the task effectively. As you monitor students' work, try to remain as unobtrusive as possible. Consider the following scenario.

> Ms. Hover has planned her lesson, and she has her students working in groups of four. She walks around the room to monitor students' work and their progress. As she approaches a new group, she leans in and says, "How is it going? Doing OK?" At this point, the conversation among students and their learning stops.

Rather than interrupting students' thinking, consider listening with your back facing the group that you would like to hear. This way, students are not aware that you are eavesdropping on their conversation, and their work can continue. You can then hear their authentic language and jot down any necessary notes that you may want to refer to later.

After anticipating and monitoring, you can intentionally select students to present their work or group's work to the whole class. The selection criteria can vary. You may ask for volunteers, but there is a risk in doing so as students may present the same strategy as a previous student. This means that valuable class time is lost. Consider intentionally selecting work that leads toward the lesson's objective. The sequence of students' presentations should be purposeful.

You may choose to begin with the most common approach and work toward a unique approach. You might focus on multiple representations as the criteria. You might begin by selecting the approach that uses concrete representations and then one that uses a table or chart and conclude with work showing symbolic representations. Figure 3.15 shows sample student work to share with the class.

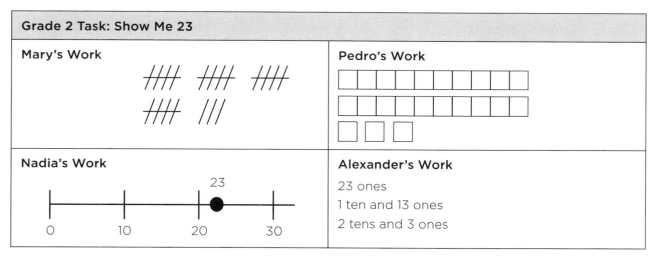

Figure 3.15: Student work to share with the class.

Mary used tallies to represent 23, while Pedro represented 23 using base-ten blocks. In terms of sharing with the class, the teacher may start with either of these two ways to show 23. Many students may have used the same approach. Nadia chose to show 23 on the number line. The teacher may select this piece of work to share next as it shows the number as a location on the number line and the student chose a scale by tens rather than counting by ones. Alexander demonstrated his knowledge by showing three compositions of 23 using tens and ones. His work might be saved for last as it may stimulate students' thinking as to how these three ways can represent the same value. Students benefit from comparing and contrasting the different approaches. In chapter 4 (page 93), we offer the Three Es strategy as a guideline for student reflection.

The order you present students' strategies in does matter, as order can deepen students' understanding as they ask questions. Asking students to compare approaches can stimulate their thinking. You can also ask students to consider which approach is the most efficient and to justify their thinking. Such questions help students *connect* with important mathematical ideas—Smith and Stein's (2011) fifth action step—and enhance classroom discussions and deepen students' understanding.

Lesson Example for Mathematical Practice 3: Pattern Possibilities

The lesson plan in figure 3.16 focuses on constructing viable arguments and critiquing the reasoning of others. In this lesson, students will identify arithmetic patterns. Although Mathematical Practice 3 is the focus of the lesson, students are also reasoning quantitatively (Mathematical Practice 2) as they solve problems (Mathematical Practice 1) and attending to precision (Mathematical Practice 6). A commentary follows the lesson providing more information related to the rationale and importance of each lesson component. Figures 3.17 and 3.18 (page 89) support the lesson's tasks.

Unit: Working with patterns and making connections between addition and multiplication

Date: April 8

Lesson: Pattern Possibilities (3.OA.9)

Learning objective: As a result of class today, students will be able to analyze patterns, explain their thinking, and justify their conclusions.

Essential Standard for Mathematical Practice: As a result of class today, students will be able to demonstrate greater proficiency in which Standard for Mathematical Practice?

Mathematical Practice 3: "Construct viable arguments and critique the reasoning of others."

- Students will analyze patterns.
- Students will draw conclusions based on those patterns.
- Students will share their justifications supporting their reasoning about the patterns and conjectures made based on the pattern.
- Students will offer feedback to each other about their conjectures.

Formative assessment process: How will students be expected to demonstrate mastery of the learning objective during in-class checks for understanding teacher feedback, and student action on that feedback?

- Students will identify patterns and then work with properties to explain and justify their thinking.
- Students will discuss their work with partners and collaborate to add more details to their responses.
- Students will identify the pattern between adding sums of consecutive numbers and representing that sum in terms of multiplication.
- Students will also write questions identifying what they are wondering about in terms of the pattern.
- Teachers will observe and monitor students' progress.

Probing Questions for Differentiation on Mathematical Tasks

Assessing Questions	Advancing Questions
(Create questions to scaffold instruction for students who are stuck during the lesson or the lesson tasks.) Task 1 • What is the meaning of **sum**? • What resource might help you? (chips to represent the groups of addends) • In the first example, one of the factors is always 3. Why?	(Create questions to further learning for students who are ready to advance beyond the learning standard.) • What do you notice about the sum and product in each case? • Can you find and explain another pattern involving products? • If we started with five consecutive numbers, what is the sum? What is the corresponding product, and what are the factors? • Explain the pattern in terms of any number of numbers.

Tasks (Tasks can vary from lesson to lesson.)	What Will the Teacher Be Doing? (How will the teacher present and then monitor student response to the task?)	What Will Students Be Doing? (How will students be actively engaged in each part of the lesson?)
Beginning-of-Class Routines How does the warm-up activity connect to students' prior knowledge, or how is it based on analysis of homework?	Place example on the board. Ask students to think independently and then turn to elbow partner. Explain why 20 × 40 = 800. "I know _____, so I also know _____." Example: I know that 2 × 4 = 8, so I know that 20 × 40 = 800. Your turn. Fill in the blanks using the same pattern as in the example: I know that ____ × ____ = ____, so I know that ____ × ____ = ____. I know that ____ × ____ = ____, so I know that ____ × ____ = ____. If students struggle, consider changing one of the factors rather than both (such as, 2 × 3 = 6, then 2 × 30 = 60). Teacher selects a few partners to share their examples and explain their thinking.	Students reflect and then share reasoning with partner. Each student fills in the sentence frames. Students then share with a partner.
Task 1 How will students be engaged in understanding the learning objective?	Review the following vocabulary: *addend*, *factor*, *sum*, *product*, *divisor*, and *quotient*. Give each student an index card with one of the vocabulary words, and tape this card on the student's back. Students should not see their word. Consider modeling effective questions for students. Suppose the vocabulary word was *difference*. "Is my word used in addition?" "Is my word an answer to an operation?"	Students ask each other questions that can be answered with a "yes" or "no" about which vocabulary word is on their back. Students determine their word. They then write an example of a number sentence and label the number that identifies their word. Example: Difference 8 – 2 = 6 6 is the difference.

continued →

Task 2 How will the task develop student sense making and reasoning? (See figure 3.17.)	Explain to students that they will be investigating patterns to a two-part task. They need to think about the pattern and then extend the pattern by writing their own examples. Discuss the three problems in question 1 (figure 3.17). Ask students to share with their partners what they notice about these examples. Look for students to notice that the addends are consecutive. Students might say that the numbers get bigger by one. The sum and product are the same. Ask, "What values would be placed in question 3d?" ($4 + 5 + 6 = 15$ $3 \times 5 = 15$)	Students work on task 1 and part 1 independently to fill in the answers to a, b, and c. They discuss with their partners what they notice and what they are wondering about. Students complete d through g and compare their work with their partners. Partners then discuss and complete numbers 4 through 6.
Task 3 How will the task require student conjectures and communication? (See figure 3.18.)	Part 2 extends students' thinking beyond three consecutive whole numbers. The sum of five consecutive numbers = 5 times the middle number. Possible questions that students might explore: Does it work with numbers that are not in order? Does it work with consecutive even numbers or consecutive odd numbers? $2 + 4 + 6 = 12$ and $4(3) = 12$ $1 + 3 + 5 + 7 + 9 = 25$ and $5(5) = 25$ Does this pattern work with an even number of numbers?	Students create examples based on Clara's thoughts. Students create at least three more examples. They then work with their partners to make a list of questions that they have and may want to explore.
Closure How will student questions and reflections be elicited in the summary of the lesson? How will students' understanding of the learning objective be determined?	Ask students to share their thoughts about the pattern. Encourage students to explain why this pattern holds true. Then ask them what they are now wondering about. Create a list of questions that students have. Encourage students to explore these questions on their own and report back to the class at a later time.	Students share their understanding about the pattern.

Source: Template adapted from Kanold, 2012c. Used with permission.

Figure 3.16: Grade 3 lesson-planning tool for Mathematical Practice 3.

*Visit **go.solution-tree.com/MathematicsatWork** for a free reproducible version of this figure.*

Task 2: Powerful Possibilities Part 1

1. Fill in the blanks.

Sum: Use the middle addend as a factor and the number of addends as the other factor. Find the product.

   a. 1 + 2 + 3 = ____   2 × 3 = ____

   b. 2 + 3 + 4 = ____   3 × 3 = ____

   c. 3 + 4 + 5 = ____   4 × 3 = ____

2. What do you notice about the sum and the products?

3. Continue the pattern:

   d. ____ + ____ + ____ = ____   ____ × ____ = ____

   e. ____ + ____ + ____ = ____   ____ × ____ = ____

   f. ____ + ____ + ____ = ____   ____ × ____ = ____

   g. ____ + ____ + ____ = ____   ____ × ____ = ____

4. What do you notice about these problems?

5. What is the connection between the addends and sum to the factors and product?

6. Is this pattern always true? Explain your thinking.

Figure 3.17: Task 2 for Mathematical Practice 3 grade 3 lesson.

*Visit **go.solution-tree.com/MathematicsatWork** for a free reproducible version of this figure.*

Task 3: Powerful Possibilities Part 2

Clara wrote this: 1 + 2 + 3 + 4 + 5 = 15   5 × 3 = 15

Create at least three more examples like Clara's example.

Explain the pattern.

Why do you think this pattern works?

What are you wondering about in terms of this pattern?

Figure 3.18: Task 3 for Mathematical Practice 3 grade 3 lesson.

*Visit **go.solution-tree.com/MathematicsatWork** for a free reproducible version of this figure.*

In the beginning-of-class routine, students explore a pattern given a fact such as $2 \times 3 = 6$, and the teacher then asks them to consider the product of 20×30. The teacher follows this by asking students to create other problems using this same pattern. Students use a sentence frame to record their work. As they share their work, they are encouraged to explain why this pattern holds true. Look for students to see that each factor is being multiplied by 10 and that, therefore, the product is 10×10 or one hundred times larger.

In the first task, students review the vocabulary that will be used in the lesson such as *addend*, *sum*, *factor*, and *product*. Using the vocabulary appropriately enhances students' ability to create effective arguments.

The second task (Powerful Possibilities Part 1) asks students to determine the connection between adding three consecutive sums and expressing the sum as a product of two factors. Students create four more examples similar to the given examples. They are then asked to think about the pattern and justify why the pattern works. Students may want to work with chips to represent the groups of addends and show how the arrangement looks different when represented as a product. Encourage students to jot down some notes responding to the questions before working with their partner. Students can then collaborate to create a logical argument as to why the connection between adding consecutive sums and expressing the sum as a product of two factors remains true.

Task 3 (Powerful Possibilities Part 2) asks students to extend their thinking beyond working with three consecutive whole numbers. The teacher asks students to evaluate whether the sum of five consecutive whole numbers can be stated as the product of two factors when following the same pattern used in working with three consecutive numbers. Be sure to emphasize how students are constructing their arguments and critiquing the reasoning of others.

Summary and Action

Standard for Mathematical Practice 3 requires students to develop cohesive arguments to justify their approaches and validate their solutions. This Mathematical Practice also requires students to develop their listening skills and ability to ask thoughtful questions. As students work collaboratively with partners or in groups of four, those who can ask clarifying and probing questions and offer summarizing statements contribute significantly to each group's progress. Mathematically proficient students go beyond correct answers; they can explain their thinking to others, and they can analyze others' thinking.

Identify a content standard you are having students learn currently or in the near future. Choose at least two of the Mathematical Practice 3 strategies from the following list to develop the habits of mind in students in order to construct viable arguments and critique the reasoning of others.

- Answers First
- What Doesn't Belong?
- Sentence frames
    - What's Under the Cup?
    - Reach 100
    - I Know This, So I Know That
- Written explanations
- Work comparisons
    - Find the Error
    - Make Suggestions
    - Factor Facts
- Peer review
- Rubrics
- Student Share

Record these in the reproducible "Strategies for Mathematical Practice 3: Construct Viable Arguments and Critique the Reasoning of Others." (Visit **go.solution-tree.com/MathematicsatWork** to download this free reproducible.) How were all students engaged when using the strategy? What was the impact on student learning? How do you know?

Chapter 4

Standard for Mathematical Practice 4: Model With Mathematics

Mathematics is a performance, a living act, a way of interpreting the world.

—JO BOALER

How many students understand that mathematics happens outside of the designated mathematics block of time each day? How do students use mathematics to make sense of their world by quantifying? For example, when are students telling time or determining how much time is left until lunch? How do they figure out how many carrot sticks they need for a party if each student gets two? How can they find the difference between the number of students who walk to school and the number who ride the bus? In other words, how do students experience the relevance of mathematics?

When working in a second-grade classroom, we overheard a student say, "I hate math. I'm never going to use this." Not twenty minutes later, the student asked a friend if he would share his apple slices with her. The friend replied, "That means I would only get three slices instead of eight." She quickly corrected him and let him know they would each get four slices to eat. This seemed acceptable to both students, and the apples were shared. Unknowingly, they had used mathematics to solve a real-world problem.

How can teachers promote students modeling with mathematics? How do you help them wonder about and make sense of their world using pictures, words, objects, and equations? Teachers can develop this habit of mind in students by using several strategies and carefully selected tasks. Figure 4.1 (page 94) shows one such task.

Interestingly, when solving the task, some students started the task by dividing $4\frac{3}{4}$ cups of flour by 2, and others multiplied the cups of flour by $\frac{1}{2}$. The student in figure 4.1 found the correct answer by multiplying the cups of flour needed for the chocolate chip cookie recipe by one-half and then adding the number of cups of flour needed to make the sugar cookies. He made sense of the task with his understanding of making cookies. However, this student made an error related to precision when he wrote "+ $3\frac{1}{2}$" at the end of the first line of work because now the expressions linking the equal signs are not all equal to one another.

Grade 5:

You want to make cookies after school. You plan to make sugar cookies and chocolate chip cookies, but you notice you do not have a lot of flour. The recipe calls for $3\frac{1}{2}$ cups of flour to make the sugar cookies and $4\frac{3}{4}$ cups of flour to make chocolate chip cookies. If you only make half of the chocolate chip cookie recipe, how many cups of flour will you need to make both batches of cookies?

Student Work

$$4\frac{3}{4} \times \frac{1}{2} = \frac{19}{8} = 2\frac{3}{8} + 3\frac{1}{2}$$

$$2\frac{3}{8} + 3\frac{4}{8} = \boxed{5\frac{7}{8} \text{ cups of flour}}$$

Figure 4.1: High-level task for Mathematical Practice 4.

*Visit **go.solution-tree.com/MathematicsatWork** for a free reproducible version of this figure.*

The fourth Standard for Mathematical Practice, "Model with mathematics," focuses on students' abilities to understand how to use mathematics to solve real-world problems that you or other students pose, or from experiences in their world. Solving the tasks will require the use of different representations and strategies as in Mathematical Practice 1: "Make sense of problems and persevere in solving them."

Students engage in this Mathematical Practice when they generate, represent, or solve problems relevant to their lives or that help make sense of the world around them. Students will be able to think of several strategies that they could use to solve real-life tasks and choose one that they feel works most effectively. Additionally, students will evaluate the reasonableness of their model and of their final solution. Through this work, students will be able to make connections between learned mathematical concepts and their own life experiences.

Table 4.1 shows more examples for what students do during a lesson to demonstrate evidence of learning Mathematical Practice 4 and what actions you can take to develop this critical thinking in students.

Understand *Why*

As students learn mathematics in grades K–5, they are learning foundational skills to apply to life experiences as well as to future college and career options. Mathematical Practice 4 invites students to think about the relevance of the mathematics learned and solve real-world problems using models such as pictures, diagrams, tables, graphs, equations, and tools.

Table 4.1: Student Evidence and Teacher Actions for Mathematical Practice 4

	Student Evidence of Learning the Practice	Teacher Actions to Engage Students
Model with mathematics.	Students: • Make a plan to solve a task prior to starting the solution pathway • Choose from multiple representations to use as the model to make sense of and solve the problem • Make assumptions and approximations to simplify a problem • Identify the critical information in the task they need to solve the problem • Make changes in the model as needed to end with a reasonable answer • Estimate an answer prior to solving the task • Identify real-world situations to which to apply mathematics	Teachers: • Model and connect different representations that students can use to solve tasks • Provide opportunities for students to share how mathematics makes sense of their daily life • Have students share models and assumptions to solve a real-world problem • Provide nonroutine, real-world problems for students to solve with mathematics models • Encourage students to estimate a solution prior to starting the task • Ask students to explain why a model was chosen and evaluate the efficiency of various models

*Visit **go.solution-tree.com/MathematicsatWork** for a free reproducible version of this table.*

In *Adding It Up: Helping Children Learn Mathematics*, Jeremy Kilpatrick and colleagues (2001) write:

> Children today are growing up in a world permeated by mathematics. The technologies used in homes, schools, and the workplace are all built on mathematical knowledge. Many educational opportunities and good jobs require high levels of mathematical expertise. Mathematical topics arise in newspaper and magazine articles, popular entertainment, and everyday conversation. (p. 15)

Students do not always recognize their mathematical world, and yet, they must be able to represent and solve real-life problems, some of which they have not even considered yet. By practicing this art of flexible thinking and application, learning in the present and for future application occurs.

One of the five interrelated strands necessary to demonstrate proficiency with mathematics in is strategic competence (Kilpatrick et al., 2001). This strand "refers to the ability to formulate mathematical problems, represent them, and solve them" (Kilpatrick et al., 2001, p. 124). Proficiency with strategic competence includes creating the question to solve, identifying what it is you need to solve the problem, clearly representing the problem and solution mathematically, and checking for the reasonableness of the final solution. This proficiency strand is a direct match to Mathematical Practice 4, "Model with mathematics."

Jo Boaler (2008), author and professor of mathematics at Stanford Graduate School of Education, suggests:

> When students try to memorize hundreds of methods, as students do in classes that use a passive approach, they find it extremely hard to use the methods in any new situations, often resulting in failure on exams as well as in life. The secret that good mathematics users know is that only a few methods need to be memorized, and that most mathematics problems can be tackled through the understanding of mathematical concepts and active problem solving. (p. 41)

In order for students to truly model with mathematics, they must develop a strong conceptual understanding of mathematics and demonstrate flexibility with choosing an effective solution pathway. It does not mean students must first be exposed to every nuance of a concept prior to applying the concept or strategy to a real-life task. Nor does it mean students must be given "steps" to solve the task at hand.

In *It's TIME: Themes and Imperatives for Mathematics Education*, the authors (NCSM, 2014) state, "Effective teachers of mathematics embed the mathematical content they are teaching in contexts to connect the mathematics to the real world" (p. 51). Timothy Kanold (2012a, 2012b, 2012c), in *Common Core Mathematics in a PLC at Work*, echoes this when he writes that students must be given the opportunity to solve real-world problems so they can determine how to ask questions and solve them. This means teachers must provide quality questions and also encourage students to create their own in order to build mathematics understanding. Interestingly, the twenty-sixth recommendation from the National Mathematics Advisory Panel (2008) also says that when students solve real-world problems they perform higher on similarly structured problems on assessments.

Mathematics is often considered a tool for applications. Scientists, engineers, artists, and athletes are but a few examples of professionals using mathematics to understand and grow in their field. They use inquiry to generate questions and then determine how to represent and solve the question posed. NCTM's (2014) *Principles to Actions* says:

> Learners should have experiences that enable them to . . . engage with challenging tasks that involve active meaning making . . . and acquire conceptual knowledge as well as procedural knowledge, so that [students] can meaningfully organize their knowledge, acquire new knowledge, and transfer and apply knowledge to new situations. (p. 9)

Again, students need to be able to pose or read a real-world problem and apply their knowledge of mathematics to its solution.

Finally, the role of multiple representations is important to students solving real-world problems. They must be able to flexibly move between pictures, diagrams, tables, graphs, equations, and mathematical tools to solve these tasks. Specifically, NCTM (2014) suggests the following:

- Students should use multiple forms of representations to make sense of and understand mathematics.

- Students should describe and justify their mathematical understanding and reasoning with drawings, diagrams, and other representations.

- Students should contextualize mathematical ideas by connecting them to real world situations.

- Students should consider the advantages or suitability of using various representations when solving problems. (p. 29)

Why should students learn to model with mathematics? It is a practice that, when students develop and use it, will help them quantify and make sense of their world. The challenge is to provide quality problems that are truly relevant to students' lives and also to teach them how to wonder about their world and be inquisitive in understanding the patterns and experiences that surround them.

Strategies for *How*

This practice requires relevant contexts for problem solving. How can students quantify their world? How can they represent real-life situations using mathematics? The strategies that follow show students how to see their world mathematically.

Life Observations

Too many students are unaware that they can apply mathematics outside school. The Life Observation strategies help students recognize when they can use mathematics to describe and wonder about real-world experiences.

Three Questions

The Three Questions strategy helps students begin to wonder about their world and ask questions about it that they can solve mathematically. Sometimes students cannot think about how their world uses mathematics without first generating some examples.

In this strategy, the teacher shows students a picture, and students work in pairs to generate three questions related to the picture to share with all students. You may want to suggest question starters such as the following.

- "I wonder if _____?"

- "I wonder how _____?"

- "How many _____?"

- "How much more _____ than _____?"

- "How much _____ altogether?"

When you first present the picture, allow students to create any type of question. For example, you show a picture of a father and son standing next to one another. Some students may want to ask non-mathematical questions.

- "Is that his dad?"

- "What did they eat for breakfast?"

Others may begin to ask questions such as the following.

- "How much taller is the dad?"

- "How much older is the dad?"

- "I wonder how heavy the father and son are altogether?"

Some additional pictures you could share for this activity are:

- Four apples, one of which has been cut into eight slices

- A person filling a rectangular prism fish tank with water

- A bedroom under construction with bare walls, holes for windows, and no flooring

 When you ask for student pairs' questions, make a list of questions that can be answered using mathematics and questions that cannot be answered using mathematics. See table 4.2 for examples. Have students help sort the two types of questions. All of the questions are good because students are making sense of the photo and learning inquiry, but they will need to begin identifying mathematics questions.

Table 4.2: Student Examples of Mathematics Questions Versus Other Questions

Mathematics Questions	Other Questions
• How much taller is the dad? • How much older is the dad? • I wonder how heavy the father and son are altogether?	• Is that the boy's dad? • What did they eat for breakfast?

For those questions that can be answered using mathematics, have students identify what they know and what they need to know in order to answer the question. Students or the teacher can generate the numbers or information needed to answer the question. Have student pairs choose a mathematics question to answer when they have the required information.

For example, to answer "How much taller is the dad?" students already know the dad is taller. They need to know the dad's and son's heights. When they determine the information they need, share with students the possible heights. These values can change based on the grade level. In the primary grades, the dad might be six feet and the son four feet. In the intermediate grades the dad might be six-feet-two inches and the son four-and-a-fourth feet tall or the son half the height of his father.

Figure 4.2 offers a sample worksheet for students to use when documenting their thinking with this strategy.

Three Questions

You can use mathematics to make sense of your world. After seeing each picture, write mathematical questions you think of first. Then, identify what you know and what you need to know to answer each question.

Picture 1

Question	What do you know?	What do you need to know?

Picture 2

Question	What do you know?	What do you need to know?

Figure 4.2: Three Questions recording sheet.

*Visit **go.solution-tree.com/MathematicsatWork** for a free reproducible version of this figure.*

When students determine what they need to know to answer the questions, they are building real-life experiences with mathematics applications. When adults want to replace the carpeting in a room or buy a new phone, they must first determine the room's dimensions or compare pricing plans for phone carriers. The information is not conveniently given to them at the instant they pose their question, as happens too often with word problems used in the classroom.

Over time, challenge students to create three questions that they can solve using mathematics. Students can write them on sticky notes and add them to chart paper with a picture placed at the top of the paper. They can solve one problem of their choice to conclude the activity and, if these posters are left up during a unit, can solve additional problems when they finish work early. You can decide whether to write needed information or values on the poster once questions are posed or wait and provide the information once a student or group of students determines the information they need to solve the task.

Another variation of this strategy is to place several pictures on chart paper (one picture per piece) and have the class take a gallery walk and write questions on each student's chart paper. Each pair or group of four students can use a different colored pen when writing questions.

This strategy helps students realize that there is still much to wonder about mathematically, and not every problem is predetermined. Students can make important observations when making sense of their world and can then ask meaningful questions.

Real-Life Examples

Once students know how to recognize mathematics in their world, have them work with their parents to identify how to use mathematics outside the classroom. Teachers and parents may both need practice with this. Challenge yourself and parents to generate a list of things they look at or do online, read, or solve in life that require an understanding of mathematics. For example, when a parent buys something online, what is the total cost of the items purchased? How long will they take to arrive? How heavy is the order? What are the cost options and considerations for delivery expenses?

Consider providing suggestions in a parent letter at the start of the year and then challenge students to bring in family examples once a week to reinforce how mathematics is useful outside of class. Eventually have students begin to independently identify uses for the mathematics they are learning.

It is important to be cautious with this strategy. When a job or career is tied to the mathematics, it allows students who already know they have no interest in that career to regard the mathematics they are learning as not relevant to their lives. For example, if a teacher or parent says that engineers or bankers use addition and subtraction with decimals, students who want to be artists may believe this means adding and subtracting decimals is not relevant to their current or future reality. As such, they may allow themselves to passively learn the concept rather than become proficient with the concept for later learning and discovery. Use careers with caution and keep the focus on actual student observations of relevant mathematics appropriate to their world.

Photographic Essay

In this strategy, students find pictures in magazines or online or take photos that show mathematics in the world. They determine a theme based on a concept, much like using pictures in a literacy collage to show key elements of a story. Students can determine their own theme or be told a theme when finding pictures. Once the collage is developed, students write one to two paragraphs explaining how the pictures in the collage show real-world examples of their concept.

Use this strategy as review at the end of a unit or grading period or more broadly as a general assignment with students studying various concepts throughout the year. As each concept becomes the focus of a unit, students can share the corresponding collage with others and each student in class can be challenged to find one more picture to add to the collage during the unit.

These collages can also be placed at stations, and students can rotate stations in groups to determine which mathematical concept the pictures represent while writing their inferences on a notecard. A whole-class discussion can follow with the collage's author sharing his or her rationale for the collection of pictures chosen.

For example, a concept in kindergarten might be recognizing and naming three-dimensional shapes. A student can find pictures of cubes, cones, cylinders, and spheres and label the parts or write a couple of sentences explaining the shapes in the pictures collected. In later grades, deeper connections can be made between shapes and architecture. In fifth grade, a student might create a collage showing pictures of cars being filled with gas or food being purchased at the store to show operations with decimals. Students' corresponding speech or writing would address how the pictures show evidence of the concept in the real world. The writing may even include some student-created problems others can solve.

Multiple Representations

Consider the different models and tools students might use to solve a relevant and real-world problem. Mathematical models include pictures, diagrams, tables, expressions, equations, and graphs. Students might also need tools such as a rule, protractor, or pattern blocks to solve the problem. See chapter 5 (page 117) for a further discussion of tools.

Consider the following grade 2 task. The teacher gives each group of students three lengths of string—one is six inches, one is eight inches, and one is twelve inches—and presents the following scenario.

> Sophia has three snakes. Each piece of string represents one of Sophia's snakes. The snakes can all be kept in the same aquarium. The length of the aquarium must be twice as long as the total length of the snakes altogether. How long does Sophia's aquarium need to be? All of the aquariums have lengths listed in inches. Use words, numbers, or pictures to show how you know your answer is correct.

To answer this task, students will need to know how to measure using inches and how to add three numbers. They will approach the task many ways, and it will benefit all students when those different representations are shared for all to see and understand as possible solution pathways that connect to one another.

Some students will use a ruler to measure each of the three pieces of string. Others will lay the strings end-to-end and measure from the start of one piece to the end of the third piece. Regardless, they will need to record their measurements. This record is the first representation of the task (see figure 4.3, page 102).

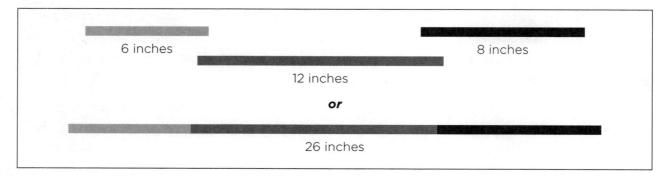

Figure 4.3: Grade 2 measurements of three pieces of string.

To solve the task, students will then need to find the sum of six inches, eight inches, and twelve inches and double that result. Some will already have the total length and only need to double that result. Others may double each snake's length and then find the total length needed for the aquarium. All of these strategies demonstrate that a student has made sense of the problem and is on an identified solution pathway.

Once students have used different models to solve the problem, have students share their solution strategies and models with one another to broaden their understanding of models they can use to solve problems and how the various models are connected (Smith & Stein, 2011).

A few possible models to represent problems in grades K–5 include:

- Ten frames
- Unifix or linking cubes
- Objects
- Drawings (for example, objects representing people or things in a word problem, area models, and geometry figures)
- Tallies
- Number lines
- Open number lines
- Base-ten pieces
- Hundreds charts
- Recording the use of tools (for example, rulers, protractors, calculators)
- Organized table of values
- Expressions
- Measurement
- Equations
- Graphs (for example, bar graph, picture graph, line plot, and coordinate plane)

The representations students need to solve the snake task involve showing how to think about finding the total length of the aquarium needed. As students are solving the problem, they will need to stop and make sure their model is leading them to a reasonable solution. Figure 4.4 offers sample student representations with a circle for the mathematical model. To reference other possible strategies for adding and subtracting, see chapter 1 (page 13).

Model	Example
Equation	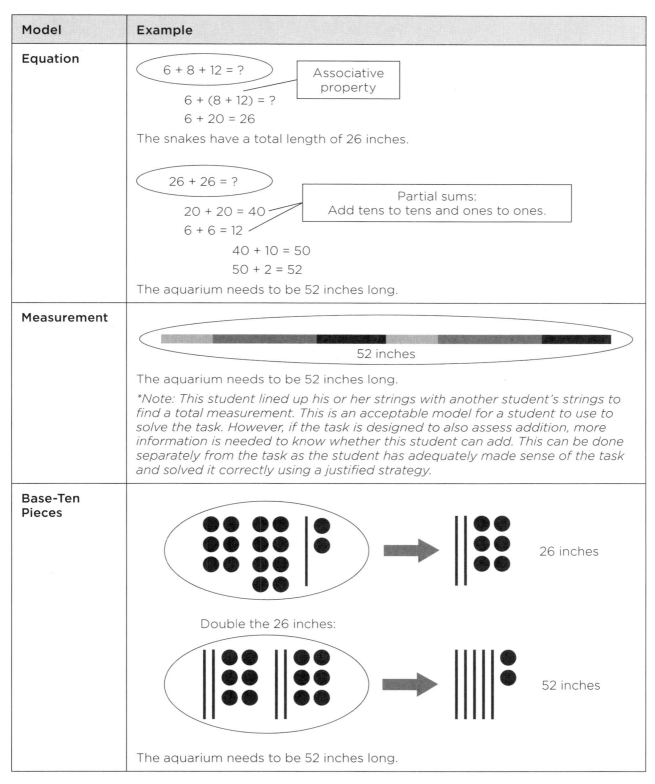
Measurement	
Base-Ten Pieces	

continued →

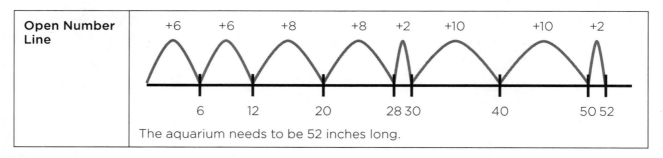

Figure 4.4: Sample student representations.

As students learn different ways to represent solution pathways and model problems, they can use a graphic organizer to help make connections between them. Consider the Rule of Five graphic organizer in figure 4.5, which stems from the five common mathematical representations: (1) graphical, (2) numerical, (3) symbolic, (4) written, and (5) contextual. Keep in mind not all grade levels will use each type of representation; this is simply an organizer showing the connections between models students might choose when solving tasks. Students can use such a graphic organizer to practice representing or solving a problem using different representations. You will need to be clear about which representations to have students show for a task before providing the graphic organizer.

Rule of Five	
Concrete (Picture)	**Table**
Graph	**Symbolic (Equation)**
Verbal or Written Summary	

Figure 4.5: Rule of Five graphic organizer.

*Visit **go.solution-tree.com/MathematicsatWork** for a free reproducible version of this figure.*

Consider the following kindergarten task.

A box holds 10 crayons. You have put 6 crayons in the box.
How many more crayons can you fit in the box?

When solving this task, younger students would not be able to show every type of representation on their Rule of Five graphic organizer. A modified graphic organizer for kindergarten might look like the one in figure 4.6. Note that students might use a ten frame, Unifix cubes, or actual crayons as examples for the object section of the template. As the teacher, you will need to see the objects students use to answer the question. Students might also draw pictures. In the case of figure 4.6, the teacher records the student's verbal summary.

Concrete (Picture)	Symbolic (Equation)
‖‖‖‖ ⟮‖‖‖⟯	10 – 6 = ?

Verbal or Written Summary
If I put 6 crayons in a box, I can add one more, one more, one more, and one more to fill the box. That means I can put 4 more crayons in the box.

Figure 4.6: Kindergarten Rule of Five example.

Consider the following third-grade task.

> Martin fills one-pound boxes of fruit to sell. He can fit 3 apples in one box, 3 oranges in one box, and 3 pears in one box. He sells 6 boxes of apples, 4 boxes of oranges, and 5 boxes of pears. How many total pieces of fruit did Martin sell?

The third-grade Rule of Five graphic organizer might look like the one in figure 4.7. Each representation leads to a solution strategy that gives the correct answer.

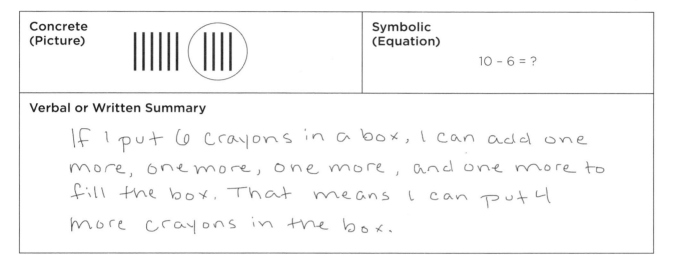

Concrete (Picture)

Table

Type	Amount of Fruit
Apple	18
Orange	12
Pear	15

continued →

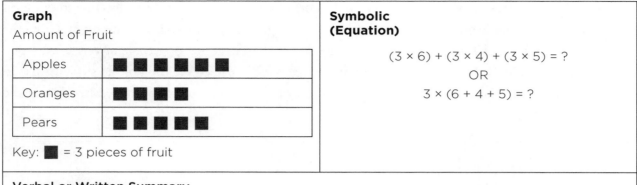

Figure 4.7: Grade 3 Rule of Five example.

Much like Mathematical Practice 1, "Make sense of problems and persevere in solving them," students need to know there are many mathematical models they can use to solve a real-world problem. If one model does not seem to produce a reasonable answer, students can use another model. Challenge students to always think of other ways to solve problems or represent solutions as they model. Challenge them to also compare one model to another by talking about what each model helps them understand or see when solving a task.

Three Es

There is never just one strategy or representation a student can use when solving a real-world task. As the class uses various models, students build a toolkit of ideas for representing problems and planning solution pathways. As their toolkit grows, students need to start choosing which strategy they believe is most productive for them to solve each task. They can use the three Es (easy, efficient, effective) for that purpose by asking themselves the following questions.

- "Is the model *easy* to use?"

- "Is the model *efficient* to use?"

- "Is the model *effective* to use?"

At first, students will not know how to answer these questions. When students are sharing their mathematical models for real-world problems in class, collect student work and ask these three questions. Have students explain what it means to be easy, efficient, and effective. Generate an anchor chart showing class understanding for these terms (see table 4.3). It is important to emphasize that what is easy for one student may not be easy for another. You may even want to create student models for the sole purpose of having students discuss which ones are easy, efficient, and effective to use.

Table 4.3: Easy, Efficient, and Effective Descriptions

Easy	The model is understood without great effort or difficulty, and when one looks at the work the computations are manageable and understandable.
	Students might say a strategy is easy for them if they can use it for many problems, it makes sense to them, it is one of the first strategies they think of when solving a problem, or they feel confident and comfortable using it.
Efficient	The model used produces a solution in a clear, concise way. There is not much wasted time and effort along the solution pathway.
	Students might say a strategy is efficient if it is a shorter way of solving a problem (with respect to time or space on a piece of paper) and always works.
Effective	The model adequately represents the problem and produces a reasonable and justifiable solution.
	Students might say a strategy is effective if, when they use it, it always leads them to a correct solution.

It is important to recognize that there is not always *one* best way to model and solve a relevant problem. Work with students to determine what types of solutions satisfy their three Es.

For example, consider the following grade 4 task and student models. Which ones are easy, efficient, and effective?

> Every student in Ms. Smith's class needs 4 marshmallows for a science experiment. There are 32 students in class, and Ms. Smith wants to have one extra marshmallow for each person in case a student makes a mistake. How many marshmallows does Ms. Smith need for the class science experiment?

Suppose Ms. Smith gave this question to her class so her students could tell her the number of marshmallows needed. Figure 4.8 (page 108) shows some possible models students might use to solve this task.

Which models produce a solution pathway that is easy, efficient, and effective?

Clara made thirty-two groups of four and then added one more to each group to make thirty-two groups of five. She then counted the total number of marshmallows needed. This is an easy strategy to make sense of and complete, though tedious. This strategy is also effective because it gives a justifiable solution pathway to the correct answer. However, it is not efficient, as much extra time is spent drawing rectangles and circles instead of concisely finding the solution, and she counted her final solution rather than using operations designed to find such an answer.

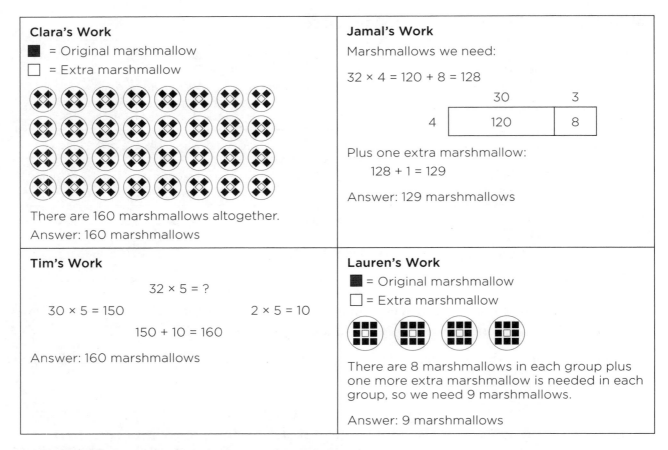

Figure 4.8: Possible student models for a grade 4 task.

Jamal solved 32 × 4 + 1. His area model for finding 32 × 4 is easy, efficient, and effective. However, his overall model is flawed because he only adds one extra marshmallow, rather than one per student. As such, his overall model is easy, could have been efficient, but is not effective as written.

Tim's work is easy to follow, though some may argue it could be made easier to follow with some words as part of the explanation. However, his model is clear—he used an equation to solve 32 × 5. He then finds partial products using the distributive property to find the product of the three tens and five and then the two ones and five before summing the partial products. His model is easy, efficient, and effective. Tim does not need the multiplication standard algorithm to make this efficient.

Lauren divided the thirty-two marshmallows into four equal groups and then added one extra marshmallow to each group. She has not made sense of the problem, so her model is ineffective. While grouping can be easy (and tedious, sometimes making it inefficient), this model is highly ineffective since it relies on division. Her answer is not reasonable.

It's important to note that the model itself does not satisfy the three Es—it is the model's relationship to the real-world problem that determines whether the model is easy, efficient, and effective. For instance, the grouping model can be effective as in the case of Clara or ineffective as in the case of Lauren.

As students create models to solve problems, have them read a partner's model and predict if it is easy, efficient, and effective. After solving the problem, ask these questions again and have students share their own work or their partner's work when they think they see a model that is easy, efficient, and effective. When seeing mathematical problems in life, they will need to determine a productive model that will lead them to their solution.

Lesson Example for Mathematical Practice 4: Addition, Subtraction, and Graphs

The first-grade lesson plan in figure 4.9 focuses on students modeling with mathematics through addition and subtraction problems, including the use of a bar graph. The real-world tasks allow for several representations, and students will demonstrate an understanding of real-world word problems based on how they solve the problems.

Although Mathematical Practice 4 is the focus of the lesson, students will also be making sense of problems and persevering in solving them (Mathematical Practice 1), talking and working together as they construct viable arguments and critique the reasoning of others (Mathematical Practice 3), attending to precision through their writing (Mathematical Practice 6), and reasoning abstractly and quantitatively (Mathematical Practice 2). A commentary follows the lesson providing more information related to the rationale and importance of each lesson component. Figures 4.10 and 4.11 (pages 112–113) support the lesson's tasks.

Unit: Addition and Subtraction

Date: January 10

Lesson: Solve real-world problems using addition and subtraction. (1.OA.1 and 1.MD.4)

Learning objective: As a result of class today, students will be able to represent and solve word problems.

Essential Standard for Mathematical Practice: As a result of class today, students will be able to demonstrate greater proficiency in which Standard for Mathematical Practice?

Mathematical Practice 4: "Model with mathematics."

- Students will represent real-world problems.
- Students will use their representations to solve real-world problems.
- Students will check for the reasonableness of their answers.

Formative assessment process: How will students be expected to demonstrate mastery of the learning objective during in-class checks for understanding teacher feedback, and student action on that feedback?

- Students will share different ways to find a missing addend, and teacher will make connections between the strategies.
- Students will create questions that can be solved using mathematics in task 1 and share them with the class. Have students critique the reasoning of others.
- Students will work in pairs to complete the bar graph in task 3, generate questions, and solve a question. The teacher will monitor progress and ask assessing and advancing questions as needed.

continued →

Probing Questions for Differentiation on Mathematical Tasks	
Assessing Questions (Create questions to scaffold instruction for students who are stuck during the lesson or the lesson tasks.) • What do you know? • What do you need to know? • How will you show your thinking? • Can you show me a model using Unifix cubes (or other objects)? • How can you use a number line to show your thinking?	**Advancing Questions** (Create questions to further learning for students who are ready to advance beyond the learning standard.) • Write another question that can be solved using the data in the bar graph in task 3. • How many more students chose apples and grapes together than chose oranges? If the number of students who chose each type of fruit is doubled, which type of fruit would be chosen the most? Explain your thinking.

Tasks (Tasks can vary from lesson to lesson.)	**What Will the Teacher Be Doing?** (How will the teacher present and then monitor student response to the task?)	**What Will Students Be Doing?** (How will students be actively engaged in each part of the lesson?)
Beginning-of-Class Routines How does the warm-up activity connect to students' prior knowledge, or how is it based on analysis of homework?	Bring students in a circle and ask: What is the value of the box in $5 + \square = 11$? How did you find the answer? Tell your elbow partner your answer and how you solved the problem. Collect answers and strategies. Show the strategies on the board for all students to understand. Ask: Which strategy makes the most sense to you? Which one are you most likely to use? Why?	Students think independently about how to find the answer and explain how to find the answer. Students share their answer and strategy with an elbow partner. Students share their answers and strategies to the whole group. Students raise their hand or stand to show which strategy they would use. Selected students verbally share their answers.
Task 1 How will students be engaged in understanding the learning objective? (See figure 4.10.)	Teacher shows a picture of a group of various kinds of pets and asks, "What questions can you ask about the picture?" Teacher asks for students to share their questions and sorts them as (1) questions that can be solved using mathematics and (2) questions that cannot be solved using mathematics. Teacher chooses one of the mathematical questions for student pairs to solve.	Students work with partners to write at least two questions related to the picture. Students share questions and help sort the questions. They talk about what they need to answer each question. For example: • How many dogs are there? (Students will need to know how many dogs are in the picture.) • How many more fish than birds are there? (Students will need to know how many fish there are and how many birds there are in the picture.) Students solve the question and share the way they represented the problem and solved it.

Task 2 How will the task develop student sense making and reasoning?	Teacher asks students to determine if they like apples, grapes, or oranges the best. They must choose one. Select one student from each category to count the number of students in his or her category. On the board are three pictures: (1) an apple, (2) grapes, and (3) an orange. Students line up in front of the picture showing their favorite of the three fruits. Teacher asks students to look at their student bar graph. Then, the teacher asks: • "Which fruit did most students choose? How do you know?" • "Which fruit did the fewest number of students choose? How do you know?"	Students choose apple, grapes, or orange. Students line up in front of the fruit they chose. The teacher decides if they stay standing or sit in their line. The selected students count the number of students in their line of students and record the number on the board under the picture. Students work with an elbow partner in their line to answer these questions. Students share their answers verbally when selected by the teacher. A second student responds with "I agree because . . ." or "I disagree because . . ."
Task 3 How will the task require student conjectures and communication? (See figure 4.11.)	Teacher has students complete a bar graph showing the data from task 2. Teacher asks student pairs to create a question that can be answered using the data in the bar graph. Look for questions that require addition or subtraction in order to be answered. Teacher decides if each pair can choose a question to answer and show work when answering it or if the entire class should answer the same question. If students finish early, they can answer another question or create a new question and answer it. Select students to share their work with the class who solved different problems or who solved the same problem using different strategies.	Students complete the bar graph in task 3 with a partner using the data on the board. Students create questions that can be answered using the data in their bar graph. Students share questions, and the teacher writes them on the board or anchor chart paper. Students write the question they are going to solve and work in pairs to show their reasoning used to solve the question. Students explain how they know their answer is reasonable. Students share their work with the class.

continued →

Closure	Teacher poses the following questions:	Students answer their questions in a journal or as a pair-share and articulate what they learned during the lesson.
How will student questions and reflections be elicited in the summary of the lesson? How will students' understanding of the learning objective be determined?	• How did you know whether or not your question could be solved using mathematics? • Which strategies shown were easy for you to understand? • Which strategies were most efficient? • Which strategies were most effective? • How did the bar graph help you answer the questions about fruit students chose?	Students tell an elbow partner what they learned today related to representing and solving word problems.

Source: Template adapted from Kanold, 2012c. Used with permission.

Figure 4.9: Grade 1 lesson-planning tool for Mathematical Practice 4.

*Visit **go.solution-tree.com/MathematicsatWork** for a free reproducible version of this figure.*

Task 1

Name: _____   Partner's name: _____

1. Look at the following picture. What questions can you and your partner ask about the picture? Can you think of questions to ask that use mathematics to find an answer?

Questions

2. Solve the question from the class. Use words, pictures, or numbers to explain how you know your answer is correct.

Figure 4.10: Task 1 for Mathematical Practice 4 grade 1 lesson.

*Visit **go.solution-tree.com/MathematicsatWork** for a free reproducible version of this figure.*

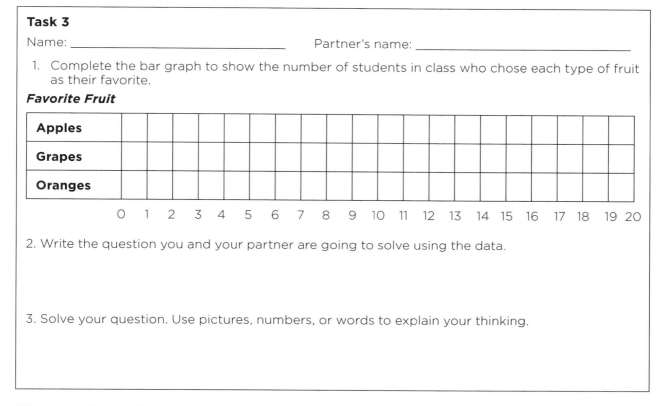

Figure 4.11: Task 3 for Mathematical Practice 4 grade 1 lesson.

Visit **go.solution-tree.com/MathematicsatWork** *for a free reproducible version of this figure.*

During the beginning-of-class routine, the teacher reminds students how to use addition or subtraction to find a missing addend. Given an equation, they will share different ways to mentally explain the answer of 6, given the questions "What is the value of the box in $5 + \Box = 11$? How did you find the answer?" Among other representations, some students will do the following.

- Students may count on and think 5, 6, 7, 8, 9, 10, 11 and realize they counted on by 6 numbers so the answer is 6.

- They may solve $11 - 5 = 6$. To do this mentally, they may first find $11 - 1 = 10$ and then $10 - 4 = 6$ using a break-apart method. Students might also find $11 - 10 = 1$ and then $1 + 5 = 6$ using another break-apart method.

- Some students may know double rules and use 10 as a double of 5, but 11 is one more than 10 so they need $5 + 1 = 6$ more.

- They could visualize a ten frame with the top row filled in. They know they still need to fill the bottom row and one more space to make 11, which means there are 6 boxes to fill in (see figure 4.12, page 114).

- Students might use their fingers to start with 5 and see that they need 6 more to make 11.

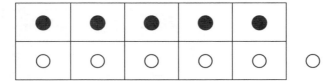

Figure 4.12: Visualized ten-frame example.

The beginning-of-class routine will connect to prior work done with addition and subtraction and spark students remembering these strategies, which they'll then apply to real-world problems posed throughout the lesson. At the end of the beginning-of-class routine, ask students which strategy they would use and why. This begins the discussion about which strategies are easy, efficient, and effective.

The first task provides an opportunity for students to see what real-life questions they can generate from a picture that requires mathematics to solve. This helps students understand the importance of information needed to solve a problem. Students will generate questions with a partner, and the teacher will sort questions as questions that can be solved using mathematics and questions that cannot be solved using mathematics. The class will then choose one question to solve, and students will determine what they need to know to complete the task with a partner. Insist that students write their thinking on a whiteboard or on paper. Select two to three students to share their solution strategies. Focus on how they first represented the tasks (for example, with equations, pictures, and objects).

The second task provides an opportunity for students to organize relevant data using a bar graph as a way to represent the data. Students will choose a favorite fruit from the choices of apples, grapes, or oranges and line up in front of a corresponding picture. Once students are in a line for the fruit chosen, the teacher selects a student from each line to count students in each category and write the number on the board. The teacher will ask some identifying mathematical questions about the bar graph like "Which fruit was chosen the most? Which fruit was chosen the least?" and have students answer.

Note that the standards (1.OA.1 and 1.MD.4) are written for sums within 20 and their inverses. If there are more than twenty students, you may need to assign additional roles like recorder and organizer, or you may need to pair students and have each pair represent 1 in the bar graph.

The third task extends task 2 and incorporates students generating questions that they can answer using mathematics. Students might want to use different colored pencils or crayons when completing the bar graph to more clearly show the number of students who chose each fruit.

During task 3, students should generate questions that require addition, subtraction, or both to solve. Examples include:

- How many more students chose apples than oranges?

- How many students chose apples and grapes?

- How many students chose apples, grapes, and oranges?

To close the lesson, students will revisit their learning and identify which strategies they have seen during class that are easy, efficient, and effective. They will revisit how a bar-graph representation can help them model real-world data and how they can ask questions that help them make sense of their world using mathematics. Focus on how students are creating and using mathematical models to solve problems.

Summary and Action

Standard for Mathematical Practice 4 requires students to recognize situations outside of class when they can use mathematics to make sense of their world. Students will begin to wonder about the application of mathematics and create mathematical models to solve the real-world problems. They will have to determine which models are best given different contexts. Throughout the practice, students will have flexibility with a variety of strategies for solving problems.

Identify a content standard you are having students learn currently or in the near future. Choose at least two of the Mathematical Practice 4 strategies from the following list to develop the habits of mind in students in order to model with mathematics.

- Life Observations
  - Three Questions
  - Real-life examples
  - Photographic essay
- Multiple representations
- Three Es

Record these in the reproducible "Strategies for Mathematical Practice 4: Model With Mathematics." (Visit **go.solution-tree.com/MathematicsatWork** to download this free reproducible.) How were all students engaged when using the strategy? What was the impact on student learning? How do you know?

Chapter 5

Standard for Mathematical Practice 5: Use Appropriate Tools Strategically

The ability to strategically select and use tools means a student is able to decide when and how to use it and, most of all, when not to.

—CATHY SEELEY

While entering an elementary school for parent conferences, a couple of bins near the door struck Sarah's eye. A sign over the enormous tubs read "Free Resources—Please Take." No further directions were provided, and parents were left to wonder why or how the resources would benefit their children's learning.

When peering into one bin, Sarah saw many three-dimensional geometric models, some fraction bars, ten frames with plastic discs, and spinners. Many of the objects, covered in a thin layer of dust, looked relatively new and unused. Why were these mathematical tools sitting in the hallway? Why were they offered free to the community? Were these extras?

It turns out the tools in the bins were simply discovered in the back of closets and classrooms that were recently renovated and cleaned. Though teachers did not have similar tools in their classrooms, they were willing to let these go. This led Sarah to wonder:

- How are mathematical tools being used in classrooms to support student understanding of mathematics?

- When are students exploring mathematics using tools?

- When are students justifying their reasoning with the use of mathematical tools?

- How are students determining when they should use a particular mathematical tool to access or solve a task?

- Would the teachers regret this decision to discard such valuable tools that could be used in their classrooms?

Any manipulative, geometric model, object, or drawing—from a pencil and paper to computer software—is a mathematical tool students can use to understand concepts, explore mathematical ideas, and verify solutions and conclusions. To teach Mathematical Practice 5, "Use appropriate tools strategically," and have students make choices regarding efficient and appropriate tools, tools will need to be available to students and tasks given that deepen student understanding through the tools' use. Figure 5.1 shows a mathematical task modified for grades 2 and 4, requiring students to use rulers, yardsticks, or tape measures. The fourth-grade task includes examples of student work.

Grade 2

If we decorate our classroom with a paper chain that goes all the way around the room, what length of chain will we need? Explain how you found your answer.

Grade 4

Suppose we decorate our classroom with a paper chain that goes all the way around the room. Each link in the paper chain has a length of 4 inches. How many links will we need?

Group A	Group B
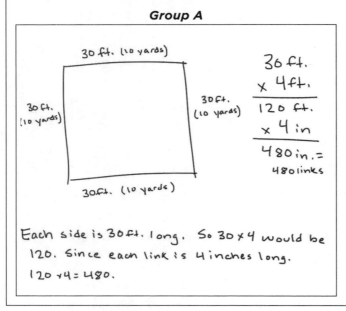	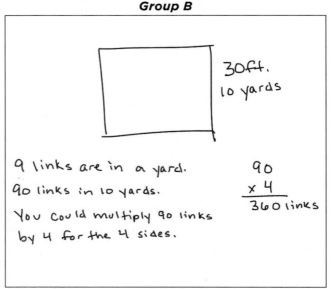

Figure 5.1: High-level tasks for Mathematical Practice 5.

*Visit **go.solution-tree.com/MathematicsatWork** for a free reproducible version of this figure.*

For the featured grade 4 task, students work in groups of four and use rulers to measure the length of the four walls in the room. Some students measure in feet using a ruler, others in yards using a yardstick. In both cases, students make decisions about which tools to use and which units to use on those tools, however, both groups do not make sense of the wall units related to the length of the links. Group A multiplies the total length of the walls (120 feet) by 4 inches, rather than realizing four-inch links meant three links per foot. Group B figures out that there were three links per foot and, therefore, nine links per yard or ninety links per ten yards. Since each wall has ten yards and there are four walls, there are a total of three hundred and sixty links needed.

As students continue to explore, understand, and create, the list of appropriate tools they might use when learning mathematics will always change. In traditional classrooms, pencil-and-paper computation made way for the modern abacus, slide rule, calculator, and computer. Which tools should students be exposed to and use today? How can students learn how to choose appropriate tools and problem solve their most effective uses in our rapidly changing world?

Students engaging in this Mathematical Practice will make choices regarding tools as well as understand how to use each one. At times, this will extend beyond tangible tools and drawings and include choices related to mental mathematics and estimation. Inherent to students developing this habit of mind is their independence and ability to make productive use of tools available when solving tasks and learning concepts, regardless of what those tools may be. Their choice will give you insights into their understanding of the mathematics they're learning.

Table 5.1 shows more examples for what students do during a lesson to demonstrate evidence of learning Mathematical Practice 5 and what actions teachers can take to develop this critical thinking in students.

Table 5.1: Student Evidence and Teacher Actions for Mathematical Practice 5

	Student Evidence of Learning the Practice	Teacher Actions to Engage Students
Use appropriate tools strategically.	Students: • Correctly use tools modeled in class • Explain why they should use a tool • Choose an effective tool to use when solving a task • Use a tool to show why a solution is correct • Explain how to use two or more tools to solve a task • Explore mathematical concepts using a tool • Make connections between concepts using tools • Compare and contrast the usefulness of tools to solve a task	Teachers: • Model the use of appropriate tools • Ask questions to prompt student reasoning about which tools to select when solving tasks or exploring mathematical concepts (for example, "How does the ruler give us the needed information? Which tool might you use to show how to solve 4 + 5 = ?") • Have tools accessible to students so students can choose the tools to use • Have students who used different tools to solve a problem share their solutions with the class and help students make connections • Ask students to identify other times they have used a tool • Develop an anchor chart with students that shows the tools students have learned to use • When a tool is new, allow students a few moments to play with the tool (especially K–2) and predict its usefulness

*Visit **go.solution-tree.com/MathematicsatWork** for a free reproducible version of this table.*

Understand *Why*

Using mathematical tools promotes deeper conceptual understanding. Tools allow students to explore and predict, make connections to previous learning, and have access to mathematical tasks so they can make sense of them and apply a solution strategy.

In *Standards for Mathematical Practice: Commentary and Elaborations for K–5*, the authors state, "Mathematically proficient elementary students choose tools that are relevant and useful to the problem at hand" (Illustrative Mathematics, 2014, p. 7). This means students will have access to and practice with many tools that promote actively participating in class. Furthermore, Kanold (2012b) notes, "In a nutshell, a variety of tools should be readily available to students to support their mathematical explorations" (p. 45).

Using mathematical tools in class allows for fun instructional experiences and can, in and of itself, engage learners. However, successful teachers have learned that using tools just for fun activities without clear links to mathematics learning does not create mathematical connections. Teachers must choose tools intentionally and purposefully for students to develop this habit of mind.

What We Know About Mathematics Teaching and Learning (McREL, 2010), a meta-study of research studies, gives another caution beyond the pitfalls of using tools for fun activities without connecting meaning, stating, "When students work with physical manipulatives, teachers must understand that the act of physically manipulating multiple pieces may create such a high cognitive load on students' thinking processes that they could lose sight of the mathematical concept being taught" (p. 13). When teachers determine students are not making a connection to mathematical content due to confusion of the tool's use, the teacher must re-evaluate that tool's effectiveness and necessity or differently scaffold its introduction to a student's toolbox.

When determining the need for students to effectively use a mathematical tool and have it as a choice in their toolbox, teachers should ask the following questions.

- "Will the tool deepen student understanding?"
- "Will the tool allow for student verification of ideas?"
- "Will the tool provide opportunities for students to explore mathematical concepts and rules?"
- "Will the tool allow students an entry point into solving higher-level cognitive tasks?"
- "Will the tool help students make sense of abstract ideas?"

When the answer to any of these is *yes*, consider the best ways to include the mathematical tool in instruction and assess a student's ability to effectively use the tool. Over time, students may find that the tool is no longer necessary as they solve and compute problems. Students' use of tools will evolve within a grade level and from one year to the next. Be careful to emphasize the link between the mathematics and the tool, rather than simply using a tool as a procedure without making such a connection. For example, when using base-ten blocks, rather than telling students how to use the blocks, have students determine that ten ones is equal to one ten block and interchange the two when solving problems.

In her book, *Smarter Than We Think*, Cathy Seeley (2014) mentions:

> As technology continues to evolve and provide us with new tools, new opportunities, and new challenges, the most lasting and transferable skill we can offer students is the ability to know when (and when not) to use an available tool. (p. 305)

Additionally, she notes, "We should never assume . . . that we know the best way for students to use a tool like a graphing calculator or a particular piece of software" (p. 306). Teachers can never model every possible tool a student might use when solving problems, nor can they model every time it is appropriate to use. Instead, teachers can model many useful tools and grow student discernment in knowing when and how a tool might be beneficial. Allow students to appropriately choose a mathematical tool and also clearly explain its connection to the task at hand.

Finally, when considering tools, take into account student access to the tool outside of mental strategies, especially in regard to socioeconomic background. How can schools provide the tools necessary to enhance and promote student learning? Armed with the knowledge for *why* and *how* students use mathematical tools, teachers must clarify grade by grade which mathematical tools *all* students must be guaranteed access to in order to improve student learning.

Strategies for *How*

The list of mathematical tools students might use to make sense of and solve tasks grows as students mature. While some tools, such as a ruler, are used at varying times throughout a student's mathematics career and life, others, such as ten frames and linking cubes used to develop an understanding of counting in the primary grades, diminish in appropriateness as mathematical concepts evolve. As described in the *Standards for Mathematical Practice: Commentary and Elaborations for K–5* (Illustrative Mathematics, 2014):

> The tools that elementary students might use include physical objects (cubes, geometric shapes, place value manipulatives, fraction bars, etc.), drawings or diagrams (number lines, tally marks, tape diagrams, arrays, tables, graphs, etc.), paper and pencil, rulers and other measuring tools, scissors, tracing paper, grid paper, virtual manipulatives or other available technologies. (p. 7)

The strategies that follow show students how to use appropriate tools strategically.

Modeling and Incorporating Mathematical Tools

It is important that students learn how to effectively use tools and connect their use to mathematical reasoning. Paper and pencil are two useful tools for every grade level. Table 5.2 (page 122) shows a list of possible mathematical tools students will appropriately use when learning content standards in the given grade bands. Know that once a tool has been used in one grade level, it remains a tool students can use when making connections and solving tasks. However, some tools will be de-emphasized since they are not as efficient and effective for students to use as the content progresses from grade to grade.

Table 5.2: Grade-Level Mathematical Tools by Domain

CCSS Domain	K–2 Mathematical Tools	3–5 Mathematical Tools
Counting and Cardinality	• Ten frames • Buttons • Unifix or linking cubes • Objects to count • Rekenreks	
Operations and Algebraic Thinking	• Number lines • Tape diagrams	• Open number lines • Coordinate planes
Number and Operations in Base Ten	• Base-ten blocks	• Base-ten blocks • Place-value charts
Number and Operations— Fractions		• Fraction bars • Fraction circles • Grid paper • Number lines • Pattern blocks • Tape diagrams
Measurement and Data	• Rulers • Bar graphs • Objects to sort • Scales • Clocks • Money	• Rulers • Cubes • Square tiles • Protractors • Scales
Geometry	• Pattern blocks • Geoboards • Tangrams • 2-D and 3-D geometric models	• Geoboards • Tangrams • 2-D and 3-D shapes

Source: NGA & CCSSO, 2010.

This chart is not an exhaustive list of all mathematical tools students might use. For example, technology, including calculators, and virtual manipulatives, are not explicitly given in the chart, yet are tools many students might use to deepen their understanding of mathematics. Additionally, students will use estimation and mental mathematics as tools in their learning. Create a list with your colleagues that clearly shows the tangible mathematical tools *all* students should have access to in each grade level to ensure vertical alignment and availability schoolwide.

It is important to model how to *use* a mathematical tool as well as how to *choose* a mathematical tool. After determining your list of tools, think about how to best model them for student use. Consider the following three steps when incorporating mathematical tools in your lessons.

1. Establish how to store mathematical tools in the classroom for future use.

2. Model the appropriate use of each tool and discuss as a class how to use it for learning. (You may need to allow time for students to explore and play with the tool.)

3. Ask questions such as:

    *   "Which tool should we use for this task? Why?"

    *   "Are there any other tools you can think of that might be useful when working on this task? If so, which tool is most effective and why?"

    *   "What information were you able to get by using this tool?"

    *   "How did this tool help you understand _____ today?"

The following three strategies are effective for modeling tools for student use: (1) introducing a tool, (2) offering student play, or (3) reviewing a tool for continued or extended use.

Introducing a Tool

Anytime you introduce a mathematical tool, make the connection to student learning clear, and restate it often. For example, state, "We are showing the number of tens and ones in a number using base-ten pieces," or "We are showing different ways to think about adding two numbers together using open number lines." Notice in both examples that the teacher gives the mathematical content and then explains the connection to the tool. Teachers can also create an anchor chart to show multiple tools to help students see connections. Later, this will promote students' ability to choose appropriate tools to solve problems. Figure 5.2 offers two sample anchor charts.

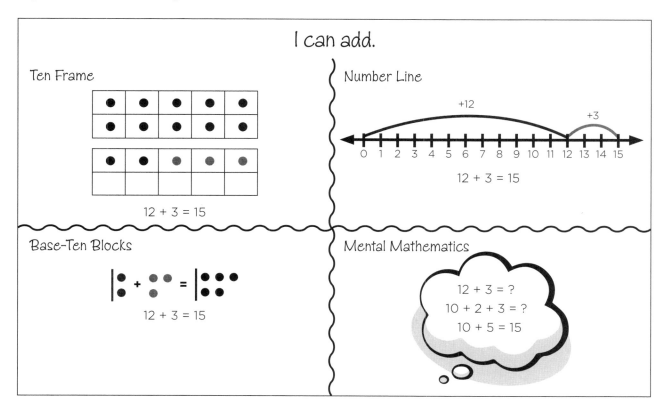

continued →

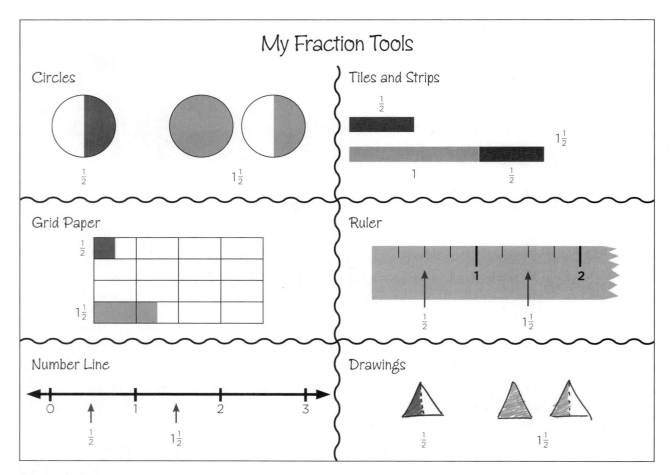

Figure 5.2: Sample anchor charts for content and tools.

As you model each tool for the first time, students will need to be able to clearly see how to use it as well as immediately practice using it with formative feedback. For example, when using a ruler, if a student does not line up the end of an object with the zero, he or she needs that feedback immediately to correctly measure and make sense of measurements. See the ruler task in chapter 6 (page 147).

Depending on the mathematical tool, you can model it first to the whole group, with students practicing at their desks, or model it in a small-group setting so students get instant feedback on their appropriate use of the tool when solving a problem. If students are at their desks, consider having them work in pairs so they can give feedback to one another.

It is critical that once you introduce a mathematical tool, students apply it to a mathematical task to begin seeing how it deepens their understanding. See table 5.3 for a few examples.

Table 5.3: Example Tasks to Promote Student Use of Mathematical Tools

Mathematical Tool	Student Task After Initial Introduction
Ten Frame (Kindergarten)	Have students work in pairs. Ask students to show different numbers on the ten frame. Have one student show a given number on the ten frame, and have the second student write the number. Students change roles for each given number.
Ruler (Grade 2)	Give each pair of students a red piece of string that is two inches long, a blue piece of string that is six inches long, a yellow piece of string that is eight inches long, and a green piece of string that is eleven inches long. Ask students to sort the string from shortest to longest and record their work using the color of the string. Next, have students alternate turns measuring the strings from shortest to longest (each student will practice measuring twice). Record the length of the string next to the color of the string already listed. How do their measurements prove their initial sorting of the string?
Geoboard (Grade 4)	Have students work in pairs, each with their own geoboard. Ask each student in a pair to create a given quadrilateral, but make sure the two quadrilaterals for the pair are not congruent. For example, tell students to create a parallelogram. One student may create a parallelogram with base lengths of 4 and a height of 3, while another creates one with base lengths of 2 and a height of 1. Ask students to draw a picture of their two shapes and then list the attributes of the shape as shown on their boards (for example, opposite sides are parallel, opposite sides have the same length).

 When modeling the tool, consider using a think-aloud, much like when modeling a reading strategy, as demonstrated in Mathematical Practice 1 and Mathematical Practice 2. Talk through what students must consider to effectively and appropriately use the tool. For example, when showing students how to use a ruler, your think-aloud might sound something like the following grade 2 example.

> "I need to line up the zero on the ruler with the end of the string because I want to begin measuring at the end. Hmmm . . . this string is curly. I think I need to straighten it out to see exactly how long it is. I know! I'll use one hand to hold one end of the string at the zero on the ruler and the other hand to stretch the string along the edge of the ruler. Oh, I see. The other end of the string ends at the six. That means the string is six inches long. I can see it stretches the full six inches from zero to six."

After students practice measuring their own pieces of string or other object, you might ask them what would happen if you lined up the first end at the 1 on the ruler instead of the 0. Have them measure the same object and explain why their end value is greater than the original measurement by one. Ask students what they can do to find the correct measurement. For example:

> "When we started at zero, the string ended at six.
> When we started at one, the string ended at seven.
> Did we get different answers for the length of the string? Explain."

Conclude by showing or having students show the following corresponding equations: $6 - 0 = 6$ inches and $7 - 1 = 6$ inches.

Although it might be easier to start the measurement at 0, it is not necessary. What is important is that students find the exact measurement and begin to make sense of how to use a ruler precisely.

In measurement, students will first explore nonstandard lengths (for example, How many paperclips long is your pencil? Which person is taller?). Standard measurement lengths often begin in second grade, but for some students, standard measurement on a ruler makes sense before nonstandard measurement, so if a student is struggling, consider teaching inches and then moving to nonstandard lengths. Measurement without gaps or overlaps is also part of Mathematical Practice 6, "Attend to precision," since placement of the measuring tool must be precise for accurate results.

Later, in the intermediate grades when students have learned about length, area, and volume, they will have to consider the attributes of the measurement needed in a task to determine the best mathematical tool. For example, students might recognize that they could use a notecard to measure the area of a rectangle rather than string because area and notecards are both two-dimensional and cover space.

Offering Student Play

In the primary grades, students need some playtime with mathematical tools before they are instructed how to formally use them. Teachers should model and explain appropriate play. When kindergartners count plastic bugs, for example, allow them to first play with the bugs for a few minutes. Next, have students sort the bugs to build their sorting abilities while noticing bugs' key features. Model by asking, "I wonder how many bugs are green?" Then, ask or show how to group the green bugs together before counting aloud. Have students do the same.

Similarly, when first introducing students to whiteboards, which allow them to share and explain their thinking, give them a few moments to familiarize themselves with the tool by allowing students to draw (appropriate) pictures on their whiteboard and erase them. Eventually, have students copy a mathematics equation to solve, and then have them show their work to practice organization and writing large enough for other students to see it when sharing. Have a model prepared in advance to show students the desired outcome, as illustrated in figure 5.3.

When using technology (including calculators) or virtual manipulatives, allow a few minutes for students to explore how the technology or tools work. Then, pose a question that invites them to explore how

Grade 1 Whiteboard Example	Grade 5 Whiteboard Example
Copy the problem on your whiteboard clearly. Write large enough for others to see your thinking when you show your board. 13 + 7 = ? 10 + 3 + 7 = ? 10 + 10 = 20	Copy the problem on your whiteboard clearly. Write large enough for others to see your thinking when you show your board. $4\frac{1}{2} + 5\frac{3}{4}$ = ? $4 + 5 + \frac{1}{2} + \frac{3}{4}$ = ? $9 + \frac{2}{4} + \frac{3}{4}$ = ? $9 + \frac{5}{4}$ = ? $9 + 1\frac{1}{4} = 10\frac{1}{4}$

Figure 5.3: Whiteboard examples.

to use the tool to find the solution before specifically modeling its use. Students need to understand what the tool might offer and explore its uses before being shown a way to use the mathematical tool effectively.

At the end of the student play, ask questions such as the following.

- "What is the purpose of this tool?"

- "How do you think we can use this tool to [add, subtract, multiply, or divide]?"

- "How is this tool like a tool you have used before?"

- "How do you think we can use this tool to solve mathematics problems?"

- "What do you like about this tool?"

- "What do you wonder about with this tool?"

Reviewing a Tool

After you have introduced a mathematical tool and students have applied the tool, review the tool with the class. When revisiting the appropriate use of a mathematical tool, consider describing the learning objective for the day and asking students which mathematical tools, if any, they can think of that might help them make sense of the tasks during the lesson. If no student identifies the tool you planned to review, ask how they could use the tool: "How might we use _____?"

Next, pose a question, and ask students how they can use the tool to complete the task. An example for grade 3 might be as follows.

Use the pattern blocks to show:
a. One-half of a hexagon—Which shape is one-half of a hexagon?
b. One-third of a hexagon—Which shape is one-third of a hexagon?
c. One-sixth of a hexagon—Which shape is one-sixth of a hexagon?

Have students explore using the tool in pairs or groups. If a student is stuck and not sure how to use the tool, have him or her talk to a partner or group. The goal is to focus on the learning objective and assume student use of the tool to reach it rather than focusing solely on the mathematical tool.

Students can make sense of fractions and operations with fractions using number lines and fraction strips. Before a student ever encounters an algorithm, provide them with fraction strips and see if they can solve the problems devising their own strategies (for examples, see figure 5.4). Fraction strips are useful for grades 3–5. Teachers should revisit them year to year to allow students an opportunity to learn through exploration.

Mathematics Problem	Solution Using Fraction Strips
Show how to decompose $\frac{3}{5}$.	Without knowing the algorithm for fraction addition, students can lay strips end to end to show: $\frac{1}{5} + \frac{1}{5} + \frac{1}{5} = \frac{3}{5}$ or $\frac{1}{5} + \frac{2}{5} = \frac{3}{5}$
Levi says $\frac{3}{8} > \frac{2}{4}$ because 3 is greater than 2. **Is he correct? Explain your reasoning.**	Students can show that three one-eighth pieces do not have a greater length end to end than two one-fourth pieces even though 3 > 2. Students can also create two number lines, one separated into eighths and the other separated into fourths, and show that $\frac{2}{4}$ is farther from 0 than $\frac{3}{8}$. Students might also reason that $\frac{2}{4}$ is equal to $\frac{1}{2}$ and $\frac{3}{8}$ is less than $\frac{1}{2}$, so Levi is incorrect.
Simplify $\frac{2}{10} + \frac{3}{10}$.	Students can line up two one-tenth pieces with three one-tenth pieces and find the sum of $\frac{5}{10}$. Some may see from comparison with the $\frac{1}{2}$ fraction piece that $\frac{5}{10} = \frac{1}{2}$.
Add $\frac{5}{6} + \frac{1}{3}$.	Students can line up five one-sixth pieces with one one-third piece. Then, they are puzzled about how to write the sum. Eventually, they will replace the $\frac{1}{3}$ with two one-sixth pieces (using one from a partner) because they will see the equivalence. They will then write the sum as $\frac{7}{6}$ or $1\frac{1}{6}$. They find a common denominator and add without an algorithm.
Multiply $5 \times \frac{2}{3}$.	Students will make five groups of two one-third pieces. When placed end to end, the product is $\frac{10}{3}$ or $3\frac{1}{3}$.
Divide $\frac{1}{2} \div 4$.	Students can look at the fraction tile chart created when $\frac{1}{2}$ is the whole and see that if $\frac{1}{2}$ is partitioned into four equal lengths, each length represents $\frac{1}{8}$. They might also think about four equal shares of half of a candy bar and determine that each share represents $\frac{1}{8}$ of the candy bar.
Divide $\frac{1}{2} \div \frac{1}{8}$.	Students can determine how many $\frac{1}{8}$ length strips equal the length of a $\frac{1}{2}$ strip. They will find the answer is 4.

Figure 5.4: Fraction strategies.

If students make their own fraction strips with paper (figure 5.5), do not label each strip with the fraction it represents. This will encourage students to refer to the whole when using the strips instead of simply reading a value on each one. Have students use colored pencils or crayons to distinctly color each row. This will help students quickly find the pieces needed when solving problems after the pieces are cut and stored in a ziplock bag.

Some students may make connections to other mathematical concepts or tools. Allow students to use other tools, and then help them and the class make connections between the tools and the solution. Which tool was more efficient? Which one did students like better? Students will see there are multiple ways to approach a solution strategy. Allow for that discussion and encourage choice as long as it efficiently and effectively deepens student understanding and verifies the solution (see the Three Es strategy in chapter 4, page 93).

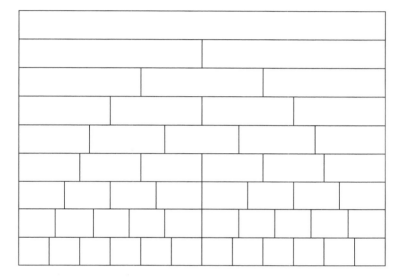

Figure 5.5: Fraction strips.

*Visit **go.solution-tree.com/MathematicsatWork** for a free reproducible version of this figure.*

Be sure to revisit tools often and make them readily accessible to students during mathematics lessons. Through practice and application, student understanding and ability to select appropriate tools increase so using mathematical tools becomes a natural way to think about and solve problems. You can use tool selection as another means of formatively assessing students' understanding of mathematics concepts.

Toolbox

A student's mathematical toolbox is made up of the mathematical tools he or she understands how to use and why. Through choosing which tool to use and applying it to tasks, students become active and engaged learners. The use of tools is never about simply teacher modeling but rather student action to make sense of mathematics. Kanold (2012a) furthers this point by stating, "This standard is not about watching the teacher demonstrating various tools. Specifically, this practice is about students experiencing the opportunity to develop an understanding by engaging in applications involving mathematics" (p. 44).

Therefore, students must have access to a toolbox of mathematical manipulatives, drawings, or technology after learning a tool. Teachers should consider how students can best access such a collection and when to remove tools to help students develop fluency in mathematical concepts. Some questions to ask yourself as you consider this are the following.

- "How can students make choices about when to use a mathematical tool?"

- "How can students quickly have access to a tool they want to use when they want to use it?"

- "How can students share their thinking with others related to their choice and application of tools?"

- "How can students gradually release the need to use some tools as mental mathematics and fluency become the effective tools and strategies?"

- "When should I remind students of mental mathematics as they develop their mathematics understanding?"

Some common strategies for distributing tools for a lesson are the following.

- Clearly mark bins with the available tools and place them on a shelf in the classroom. Allow students to take manipulatives from the common bins as needed and replace them when finished.

- Have students place mathematical tools in a ziplock bag or pencil box to be put on their desks during a mathematics lesson. Students can sort through their own manipulatives and choose the tool.

- Some tools require drawings, so students may want a mathematics toolbox journal that shows different drawings and diagrams to use as well as lists tangible mathematical tools, how they are useful, and records when students have used them to solve problems.

Each time students choose a mathematical tool, have them share with a partner, group, or class the reason the tool was chosen and how it deepened their understanding of the task or allowed them to solve the task or verify their solution. Allow all students to see how they can use tools to make sense of mathematics. If students are not used to choosing the tools they will use, consider posing a task and asking which tools they might use to solve it. You might also want to assign each group of students a different tool to use to solve the task and then have groups show how they could have chosen and appropriately used several tools. This will help students discern when it is best to use each mathematical tool. See the following kindergarten and grade 5 examples for sample classroom tasks that offer a variety of tools for students to choose from.

Kindergarten

There were some bunnies in a field. Then, seven more bunnies hopped over. Now there are ten bunnies. How many bunnies started in the field?

Possible tools students might use:

- Ten frames

- Number lines

- Base-ten blocks

- Unifix or linking cubes or other counting objects

- Paper-and-pencil computations

- Tallies

Note that some students may not be able to think abstractly and may actually need pictures of bunnies or objects looking like bunnies to solve this task. Place the tools in bins strategically around the room or at stations for students to choose from since there are several manipulatives in each bin that would exceed a personal toolbox. Again, the tools students select also provide formative feedback. Allow students to share their thinking with others. For example, have a student who used a ten frame share with a student who used a number line how his or her tool solved the problem. Consider sharing different solutions with the entire class at the carpet. Make connections between the tools and ask students how their tools helped them make sense of the problem.

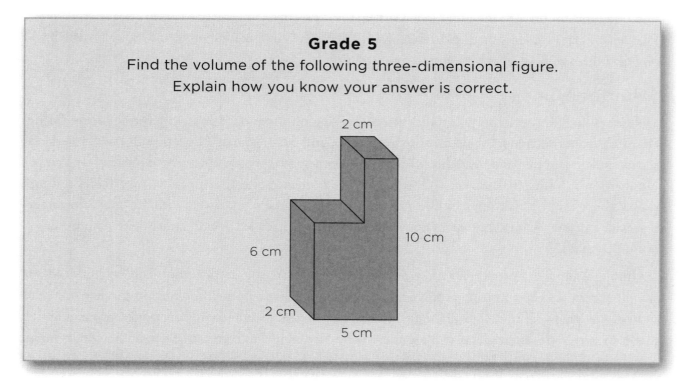

Grade 5
Find the volume of the following three-dimensional figure.
Explain how you know your answer is correct.

Possible tools students might use include the following.

- Cubes to build the figure and to determine the number of cubic centimeters that describe its volume

- Isometric graph paper to draw this three-dimensional model, find the length of missing edges, and compute the volume

- Graph paper to draw this three-dimensional model, find the length of the missing edges, and compute the volume

- Rectangular prisms to build the shape (not to scale), make sense of the shape, and then find its volume

- Paper to draw a side view, top view, and front view to determine missing dimensions and then compute the volume

- Two rectangular prisms to decompose the figure and find the sum of the volume of each prism

Regardless of whether students choose one of these tools to solve for volume, they must accurately find and justify the volume, which the mathematical tools should support.

In the intermediate grades, tools like rulers, protractors, pencils, erasers, and calculators can be stored in a personal toolbox while graph paper, base-ten blocks, whiteboards, or larger manipulatives can be stored for access around the room.

One mathematical tool that a student always has access to in his or her toolbox is the ability to compute and estimate mentally. This ability is strengthened through the use of manipulatives and, alternately, students can use manipulatives to justify or prove a mental computation. Allow students to estimate the solution to a problem prior to solving it (see Mathematical Practice 1) or verify solutions (see Mathematical Practice 6). Have students share the different ways they solve computation problems mentally, which will provide others with insights into mental mathematics.

Number Lines

Number lines are an important tool when organizing information that seems abstract or confusing. Research from the Freudenthal Institute for Science and Mathematics Education in the Netherlands shows student learning and flexibility with mental mathematics within 100 were improved using empty number lines (Klein, Beishuizen, & Treffers, 1998). Students' effective use of this mathematical tool extends beyond the classic number line introduced in the primary grades that includes only whole numbers with a tick mark for every value. Empty number lines, also called open number lines, do not contain every tick mark.

Although number lines are not often explicitly part of state standards until second grade, students should experience them as early as kindergarten as a tool to have in their toolbox. In kindergarten and most of first grade, number lines are sometimes called *number tracks* and will have whole numbers as tick marks or within boxes. Students will use them as a visual model to count, understand how to compose and decompose numbers, solve addition and subtraction problems, and deepen place-value understanding. The amount students add or subtract should be labeled above the arc showing each jump on a number line.

Number lines are also closely connected to measurement because the numbers along the edge of the ruler are a number line themselves. Later, in grades 3–5, the ruler can still be connected to fractions on the number line as students identify halves, fourths, and eighths and explore sixteenths.

Additionally, classrooms can use a human number line to make this clearer for students. Place masking tape or painter's tape on the floor to create a life-sized number line for students to stand on and walk on to discover and show a solution. As students act out the task, model how to record their thinking on a written number line.

Number Lines With Counting and Rounding

When teaching students to count in kindergarten, show a large number line to 10 on the floor (later extend it to 20). Make sure there is a tick mark for each number. Point to the 0 as the starting point and then begin chanting "1, 2, 3, 4, 5, 6, 7, 8, 9, 10." This will introduce students to a number line, allow them to see each number counted, and practice counting. You can ask students if there are numbers larger than 10. This connects to the arrow to the right of 10 on the number line.

Later, ask students questions like, "What is 1 more than 8?" They can use the number line to count on from 8 to 9. You might also ask, "What is 1 less than 5?" Similarly, students can point to 5 and move left one tick mark to 4.

Number lines also help students learn to round numbers (for example, see figure 5.6). When having third graders round 42 to the nearest ten, ask students, "Which two tens are closest to 42 on the number line?" Follow this question with, "Which of those tens is 42 closest to?" Repeat with several more problems, and have students work in pairs to round the numbers to the nearest 10 (or hundred or other value). There is no need to use a rhyme that tells students how to round up or round down based on the digit in the ones place. This becomes confusing for students. Simply have them think about a number line and visualize the answer. This reasoning for rounding will later help students round to the nearest benchmark fraction, too.

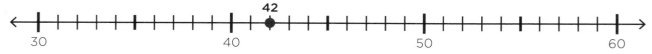

Figure 5.6: Rounding 42 to the nearest ten using a number line.

One problem that quickly arises with rounding is which 10 is the 5 (such as with 45) closest to? To answer this question, make a list of all the digits starting at 0: 0, 1, 2, 3, 4, 5, 6, 7, 8, and 9. These are the only possible digits for the ones place in any number. Orchestrate a class discussion and agreement that the digits should be evenly split, in which case, numbers that have 0, 1, 2, 3, or 4 in the ones place round to the nearest 10 less than or equal to the original number. Numbers that have 5, 6 7, 8, or 9 in the ones place round to the nearest 10 greater than the original number. See figure 5.7 for a visual of this process.

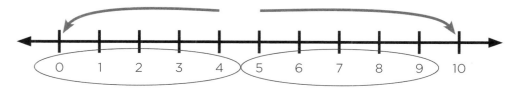

Figure 5.7: Number line showing rounding rules.

Number Lines With Operations

When building student connections with operations and numbers using number lines, model the use of the number line so students can show their thinking. Be sure the number line has arrows on both ends and shows the value of each tick mark. The examples in figure 5.8 show that the number lines can include tick marks (kindergarten and grade 1) or be open as numbers increase in size. Figure 5.8 shows tasks from various grade levels and how a student might use a number line as an effective tool to solve each.

The open number lines in figure 5.8 for grades 2–4 also help students develop the use of the mathematical strategy that involves mental mathematics. Through the visual representation, students learn to flexibly compose and decompose numbers, allowing them to access and effectively use mental computations as a mathematical tool as their skills develop.

Number Line Solution Strategy

K–1 addition: Sandi has 3 candies. She gets 4 more candies. How many candies does Sandi have now? Show your thinking.

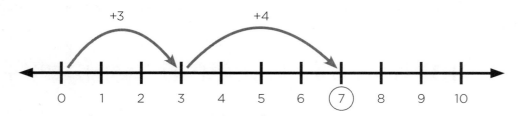

K–1 subtraction: Sandi has 7 candies now. She ate 4 candies. How many candies does Sandi have left? Show your thinking.

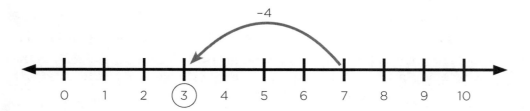

Grade 2: Jerome has a jar full of three different colors of jelly beans. There are 72 jelly beans in the jar. He put 24 red jelly beans in the jar and 35 green jelly beans in the jar. How many yellow jelly beans are in the jar? Show your thinking.

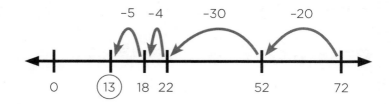

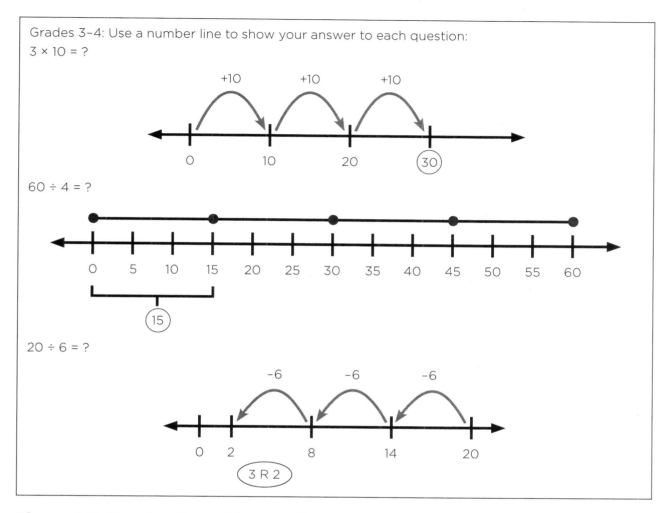

Figure 5.8: Number line with operations tasks.

Number Lines With Fractions

In addition to developing an understanding of counting and operations, students can use number lines to make sense of fractions. Consider using double number lines to show equivalence and compare rational numbers (see figure 5.9, page 136).

Number Lines With Measurement and Data

In grades 3–5, students will use number lines to show elapsed time and to organize and record measurement data in line plots. Students will also use a line plot to answer questions related to the measurements it shows. Have students connect a line plot to number lines to deepen their understanding of its use and purpose. The grade 5 task example in figure 5.10 (page 136) gives a line plot. Each *x* represents one student who walked a given distance.

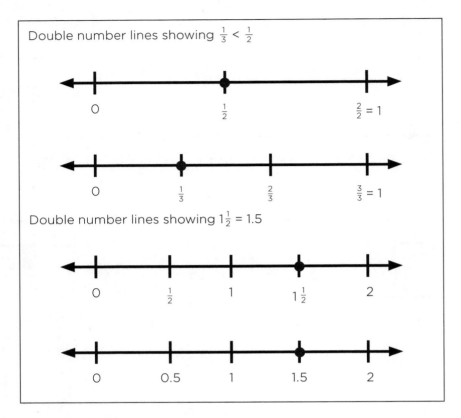

Figure 5.9: Double number line examples.

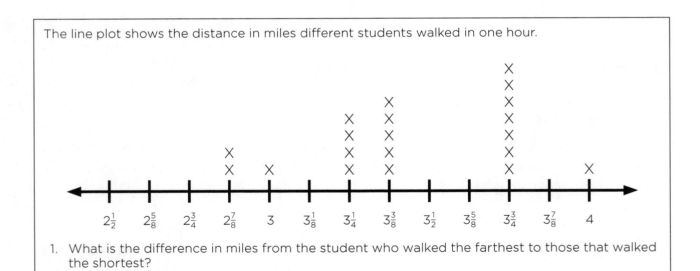

1. What is the difference in miles from the student who walked the farthest to those that walked the shortest?

2. How many students walked farther than 3.5 miles?

3. Altogether, how many miles did students walk who each walked $3\frac{3}{4}$ miles?

Figure 5.10: Line plot task.

*Visit **go.solution-tree.com/MathematicsatWork** for a free reproducible version of this figure.*

Students will have to read the line plot and understand the meaning of each x to answer the questions. They might answer question 3 by drawing a traditional number line partitioned into lengths of $\frac{3}{4}$ or an open number line, and then use the number line to show seven jumps (for the seven students) of $3\frac{3}{4}$ to find the total distance walked.

Number lines are essential mathematical tools to use with students. They build conceptual understanding and help students make connections between concepts taught in one grade level to the next. They are mathematical tools that allow students to justify answers and make sense of the tasks they are asked to solve and the numbers involved.

Later, in grade 6, students will use a number line to show all of the solutions to inequalities such as $x < 5$. In even later grades, students will solve algebraic inequalities and graph them on number lines as well as coordinate planes. Grades K–5 begin students' understanding of this effective tool.

Technology

The most commonly used technology in the classroom is a calculator. While many state assessments do not allow students to use them, students can use calculators in class to deepen understanding through exploration or as an alternate way to compute or verify solutions.

When introducing multiplication, calculators can facilitate students exploring and investigating relationships between repeated addition and multiplication. Students can type 3 + 3 + 3 + 3 + 3 and 5 × 3 to make a conjecture about the relationship between the two. Calculators can also be used to highlight connections between computations and information shown in arrays, area models, or other mathematical representations. Additional discovery calculator ideas are shared in Mathematical Practice 6. Students may also use calculators for computation when addressing word problems and for checking solutions, as these instances do not draw attention away from the learning objective but provide opportunities for immediate feedback so students will know whether they need to modify their thinking or work.

Many other mathematical tools students use in the classroom are also available electronically on websites and apps. Examples include apps for protractors to measure angles and geoboards to explore the attributes of shapes. Similarly, balance scales are available to show two sides of an equation when two sides are equal. Additionally, there are many apps and tools available to show counting, operations (addition, subtraction, multiplication, and division of whole numbers, fractions, and decimals), and properties (associative, commutative, and distributive). For instance, one app, Number Rack, targets early elementary grades and models a rekenrek to allow students to represent varying number relationships. These examples each scratch the surface of the possible electronic tools students might use to solve problems.

The National Library of Virtual Manipulatives (http://nlvm.usu.edu/en/nav/vlibrary.html) is a resource free to teachers and students filled with various virtual manipulatives. The National Council of Teachers of Mathematics (NCTM; www.nctm.org/coremathtools) also has a free resource called Core Math Tools that provides access to tools such as spreadsheets, calculators, geometry software, probability simulations, and statistical graphs, although it's often most useful at the secondary level. Its plane and coordinate geometry features are available for use in the elementary grades. (Visit **go.solution-tree .com/MathematicsatWork** to access live links.)

There is no way you can prepare students for every possible virtual manipulative or tool they may encounter in their lifetime. However, you can provide students with a foundation for approaching these by assigning technology tool tasks in which students ask and answer questions such as the following.

- "How will this tool help me understand the task?"

- "How can this tool help me solve the task?"

- "What other tools could we use to solve the task? Is this the most effective and efficient tool to use?"

The use of technology in class is typically met with enthusiasm from students. Be careful that when using this resource as a valuable and effective mathematical tool, students are connecting its use to mathematics learning. As with all mathematical tools, technology should not be used solely for fun but rather always serve an intentional purpose to improve student learning.

Lesson Example for Mathematical Practice 5: Adding and Subtracting With Tools

The lesson plan in figure 5.11 focuses on using tools strategically. The tasks show how to use mathematical tools when teaching addition and subtraction in kindergarten. Prior to this lesson, students have played with and been introduced to working with cubes and other counting objects as well as number lines and ten frames, which were all used with counting. Students will practice using different tools throughout the lesson, both in a whole group and at stations. At the end of the lesson, they will share with a partner which one worked best for them and why. This allows students to begin thinking about when it is best to use each tool in their toolbox. Although Mathematical Practice 5 is the focus of the lesson, students will need to make sense of problems and persevere in solving them (Mathematical Practice 1), construct viable arguments and critique the reasoning of others (Mathematical Practice 3), and attend to precision with each tool (Mathematical Practice 6). A commentary follows the lesson providing more information related to the rationale and importance of each lesson component. Figures 5.12–5.16 (pages 141–145) support the lesson's tasks.

Unit: Addition and subtraction

Date: March 2

Lesson: Solve addition and subtraction problems using tools (K.OA.2).

Learning objective: As a result of class today, students will be able to add and subtract using the following.

- Cubes
- Number lines
- Ten frames
- Dice
- Drawings
- Equations

(This is a list that grows as students use various tools and strategies to solve addition and subtraction problems.)

Essential Standard for Mathematical Practice: As a result of class today, students will be able to demonstrate greater proficiency in which Standard for Mathematical Practice?

Mathematical Practice 5: "Use appropriate tools strategically."

- Students will add using dice.
- Students will subtract using pom-poms and cups.
- Students will compose numbers to 10 and write addition equations using ten frames.
- Students will subtract using a number line.
- Students will add and subtract word problems using chosen tools.
- Students will check the work of others using mathematical tools.

Formative assessment process: How will students be expected to demonstrate mastery of the learning objective during in-class checks for understanding teacher feedback and student action on that feedback?

- Students will make thinking visible using objects, drawings, and equations. This will allow the teacher to give meaningful descriptive feedback to students when they are working.
- Students will share their thinking with others and to the class.

Probing Questions for Differentiation on Mathematical Tasks

Assessing Questions	**Advancing Questions**
(Create questions to scaffold instruction for students who are stuck during the lesson or the lesson tasks.)	(Create questions to further learning for students who are ready to advance beyond the learning standard.)
<ul><li>What do you know?</li><li>What do you need to know?</li><li>How will you show your thinking?</li><li>Can you act out this problem so it makes sense?</li><li>Who might help you in class?</li></ul>	<ul><li>Think of a word problem you must solve using addition that has an answer of 7. Write the problem or draw a picture to show the problem.</li><li>Think of a word problem that you must solve using subtraction that has an answer of 5. Write the problem or draw a picture to show the problem.</li><li>Roll three number cubes, and find their sum.</li></ul>

Tasks (Tasks can vary from lesson to lesson.)	**What Will the Teacher Be Doing?** (How will the teacher present and then monitor student response to the task?)	**What Will Students Be Doing?** (How will students be actively engaged in each part of the lesson?)
Beginning-of-Class Routines How does the warm-up activity connect to students' prior knowledge, or how is it based on analysis of homework?	Have students sit in a circle and ask: "What is 5 + 3? When you have an answer and can explain how to do it, put one finger on your chest. If you can think of two ways to find the answer, put two fingers on your chest [and so on]." Randomly call on students to explain how they thought about 5 + 3. Record the student answers and strategies on the board or chart paper for all students to see. Emphasize the connections between the strategies. Ask: "Which strategy makes the most sense to you? Which one are you most likely to use? Why?" Repeat with 6 – 4 = ?.	Students think independently about the answer to 5 + 3 and how they can explain the answer. They place a thumb or appropriate number of fingers on their chest. Students share their answers and strategies to the whole group. Students raise their hand or stand to show which strategy they would use. Selected students verbally share their answers.

continued →

Task 1		
Task 1 How will students be engaged in understanding the learning target? (See figures 5.12, 5.13, 5.14, and 5.15.)	Modeling the stations has occurred in prior lessons. Have students help you review how to use the mathematical tools at the following four stations. Students will spend about ten minutes at each station. Play music or have students sing a song as they move from one station to the next. One station is a minilesson station with the teacher and is explained in task 2. **Stations** 1. Add the Dice: Each student at this station needs two dice, a recording sheet, and a pencil. 2. Cup Subtraction: Each student needs ten pom-poms, a plastic cup labeled "Inside" and "Outside," and a recording sheet. 3. Make 10: Students need a deck of cards 1–9, a ten frame with plastic markers, a recording sheet, two crayons, and a pencil. 4. Subtract 2: Students need cards numbered 2–10, a recording sheet, and a pencil.	Students tell the class an expectation until they have all been shared (for example, do the work, listen, be nice to others, write your thinking). **Stations** 1. Add the Dice: Students roll two dice and find and record the sum. 2. Cup Subtraction: Students hold ten pom-poms above a cup and drop them. They create a subtraction equation: 10 – pom-poms outside the cup = pom-poms inside the cup. They check their answer. 3. Make 10: Students pick a card from a deck of cards 1–9. They place that many plastic markers on their ten frame and then fill in the ten frame to make 10. On their recording sheet they use two colors of crayons—one to draw the initial number of plastic markers on the ten frame and a second to show the needed addend. They use a pencil to write each equation. 4. Subtract 2: The student draws a card and places a dot on the number line on the recording sheet at that location. The student shows jumping left 2 to show "–2" and circles the answer. The student records the equation.
Task 2 How will the task develop student sense making and reasoning? How will the task require student conjectures and communication? (See figure 5.16.)	Teacher works with a handful of students at a table or other location. This will be a station for word problems, and the teacher will observe to see which mathematical tools students use to solve each problem. Have mathematical tools available, and record the tools students use. Pose a word problem, and record how students solve the problem. Have each student share his or her thinking while the other students listen. Make connections between strategies, and ask students which tools or drawings helped them understand and solve the problem. Repeat with a second problem if time allows.	Students use mathematical tools, talk with one another, and determine how to solve a word problem. They record their thinking. Students share their answers verbally when selected by the teacher. A second student responds with "I agree because . . ." or "I disagree because . . ."

Closure	Teacher poses the following questions and listens to student responses.	Students answer the questions with a partner and then share their responses (when randomly called) with the class.
How will student questions and reflections be elicited in the summary of the lesson? How will students' understanding of the learning target be determined?	• Which mathematical tool helped you add and subtract today? Why? • How can you use a number line to subtract? • Can you use a ten frame to solve a word problem? How? • How many ways did you find to make 10 today? • What did you learn today?	

Source: Template adapted from Kanold, 2012c. Used with permission.

Figure 5.11: Kindergarten lesson-planning tool for Mathematical Practice 5.

*Visit **go.solution-tree.com/MathematicsatWork** for a free reproducible version of this figure.*

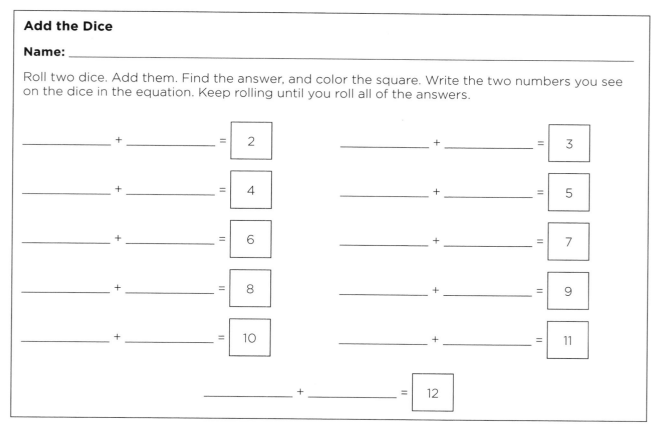

Add the Dice

Name: _____

Roll two dice. Add them. Find the answer, and color the square. Write the two numbers you see on the dice in the equation. Keep rolling until you roll all of the answers.

_____ + _____ = [2]   _____ + _____ = [3]

_____ + _____ = [4]   _____ + _____ = [5]

_____ + _____ = [6]   _____ + _____ = [7]

_____ + _____ = [8]   _____ + _____ = [9]

_____ + _____ = [10]   _____ + _____ = [11]

_____ + _____ = [12]

Figure 5.12: Task 1 for Mathematical Practice 5 kindergarten lesson—Add the Dice.

*Visit **go.solution-tree.com/MathematicsatWork** for a free reproducible version of this figure.*

Cup Subtraction

Name: _____

Drop 10 pom-poms over an empty plastic cup. Write how many fall outside the cup. Make a subtraction equation to find the number of pom-poms inside the cup. Is your equation correct? Check the number of pom-poms in the cup.

               Outside   Inside

$$10 - \boxed{} = \boxed{}$$

$$10 - \boxed{} = \boxed{}$$

$$10 - \boxed{} = \boxed{}$$

$$10 - \boxed{} = \boxed{}$$

$$10 - \boxed{} = \boxed{}$$

$$10 - \boxed{} = \boxed{}$$

Figure 5.13: Task 1 for Mathematical Practice 5 kindergarten lesson—Cup Subtraction.

Visit **go.solution-tree.com/MathematicsatWork** *for a free reproducible version of this figure.*

Make 10

Name: _____

Draw a card. Use a crayon to draw dots on the ten frame that match your number. Use another crayon to fill in the ten frame and make 10. Write the addition equation on your ten frame.

_____ + _____ = _____

_____ + _____ = _____

_____ + _____ = _____

_____ + _____ = _____

_____ + _____ = _____

_____ + _____ = _____

Figure 5.14: Task 1 for Mathematical Practice 5 kindergarten lesson—Make 10.

*Visit **go.solution-tree.com/MathematicsatWork** for a free reproducible version of this figure.*

Subtract 2

Name: _____

Draw a card. Mark the number on the number line with a dot, and show 2 hops to subtract 2. Circle the answer on the number line. Write your subtraction equation next to the number line.

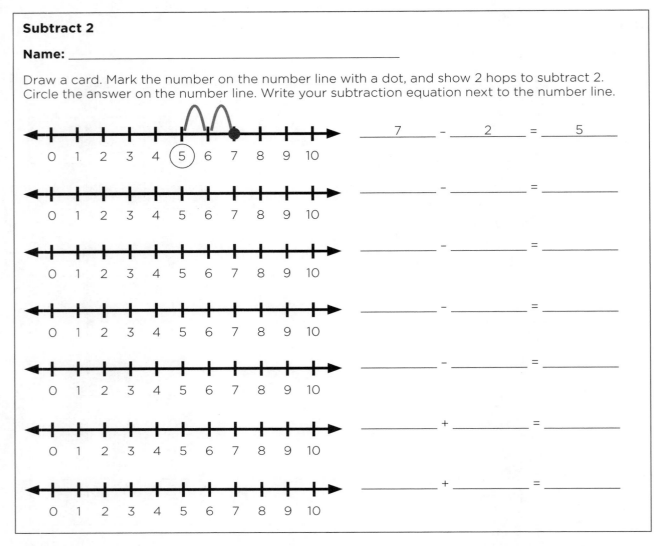

Figure 5.15: Task 1 for Mathematical Practice 5 kindergarten lesson—Subtract 2.

*Visit **go.solution-tree.com/MathematicsatWork** for a free reproducible version of this figure.*

Word Problem   **Name:** _____

> There are 3 pencils on a desk.
> Sam puts 4 more pencils on the desk.
> How many pencils are on the desk? Show your thinking.

Word Problem   **Name:** _____

> There are 2 ants on the patio.
> More ants came. Now there are 8 ants on the patio.
> How many ants came to the patio? Show your thinking.

Addition and Subtraction Word Problem Recording Sheet

Name	Tool	Tool	Tool	Tool

Figure 5.16: Task 2 for Mathematical Practice 5 kindergarten lesson.

Use the beginning-of-class routine as an opportunity to also review the stations students will visit during this lesson. Students may talk about starting with 5 and counting on 3 or starting with 3 and counting on 5. They may talk about using the doubles 3 + 3 and then adding 2 more or 5 + 5 and then subtracting 2. Some students may show a number line they used to add, and others may visualize a ten frame or use fingers. Record the thinking so all students can see the many ways to add. Stations will include adding and subtracting using ten frames, dice, number lines, and cubes. Show how to use each tool to find 5 + 3 if students do not mention these tools in their strategies, and let them know they will use these tools at stations.

The majority of time will be spent with students working in stations. Ideally, you should introduce and practice these stations in previous lessons. Consider playing music as the cue for students to clean up quickly and move to the next station. You may also want to have students sing a song, and when the song finishes, students are at their next seat. The music or songs are short to quickly move students. Each station reinforces the concept of addition and subtraction with a minilesson station with the teacher to apply this learning to solving word problems.

At the end of the lesson, have students clean their final stations and sit in a designated area. Ask them questions to connect the tools and strategies they used to solve addition and subtraction problems. Have students answer questions with a partner, and listen to see what students think they learned during the lesson. You can revisit these stations throughout the unit for additional practice as needed.

Summary and Action

Standard for Mathematical Practice 5 requires students to learn how to use tools strategically. These tools may change from year to year, and students must be prepared to learn new mathematical tools that they can use to solve problems. Most importantly, students who develop this habit of mind discern when it is appropriate and effective to use a mathematical tool and choose the best tool to use. Tools range from mental mathematics and estimation to drawings to tangible and virtual manipulatives. It is critical that students always connect the use of a tool to mathematics learning.

Identify a content standard you are having students learn currently or in the near future. Choose at least two of the Mathematical Practice 5 strategies from the following list to develop the habits of mind in students in order to use appropriate tools strategically.

- Modeling and incorporating mathematical tools
  - Introducing a tool
  - Offering student play
  - Reviewing a tool
- Toolbox
  - Number lines
  - Technology

Record these in the reproducible "Strategies for Mathematical Practice 5: Use Appropriate Tools Strategically." (Visit **go.solution-tree.com/MathematicsatWork** to download this free reproducible.) How were all students engaged when using the strategy? What was the impact on student learning? How do you know?

Chapter 6

Standard for Mathematical Practice 6:
Attend to Precision

Language is as important to learning mathematics as
it is to learning to read.

—NATIONAL COUNCIL OF TEACHERS OF MATHEMATICS

"How many minutes old are you?" a fifth-grade teacher asks her students. The question seems easy enough until students begin discussing what they need to know in order to answer this question. Working in small groups, students begin identifying some facts that they need, such as the number of days in the year, the number of hours in a day, the number of minutes in each hour, and their exact date and time of birth.

In the fourth-grade classroom down the hall, students are given a different question: "Thirteen candy bars are to be shared with seven friends. What is the size of each friend's share of the candy bars?" Two students in the class have the following dialogue.

Alfredo: Let's use a calculator. We just divide . . . 7 ÷ 13.

Alicia: Don't you mean 13 ÷ 7? We need 7 shares.

Alfredo: Right. . . . $\frac{13}{7}$ ≈ 1.857142857.

Alicia: Wow! Do we need all those digits?

Alfredo: No. Those numbers don't make sense. Looks like each person will get a little less than two candy bars.

In both of these scenarios, students attend to precision. They are using the situational context to determine how precise they need to be in their work and answer. These actions demonstrate Standard for Mathematical Practice 6, "Attend to precision."

When students attend to precision, they do so as they begin to make sense of the task presented. Students attend to precision in their own thinking and as they interact with others. Specifically, students develop this habit of mind in the areas of communication, measurement, and calculation. See figure 6.1 (page 148).

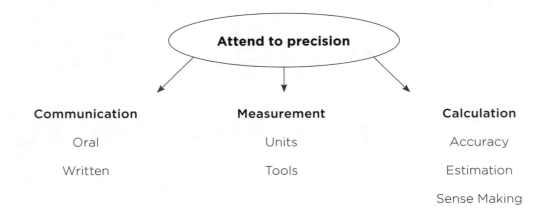

Figure 6.1: Mathematical Practice 6 categories.

As students become more mathematically proficient, their ability to communicate both orally and in writing improves. When using mathematical vocabulary, students articulate their thinking and can share their ideas clearly with others.

In a kindergarten class, we overhead the teacher ask, "How do you know that 9 is greater than 7? Remember to start your answer saying, 'I know 9 is greater than 7 because . . .'" Naturally, students' responses were intriguing.

> **Melinda:** I know because if I line up chips, one group has 7 and the other has 9. The 9 group has more in it.
>
> **Celina:** I know that 9 is greater than 7 because 9 is farther on the number line. See? (She points to the right as she faces the number line.) Any number out there is bigger than ones here.
>
> **Roger:** I know that 9 is bigger because when I count, 9 comes after 7. Each count is one more.

These students communicated and articulated their ideas clearly. They presented their thoughts with precision—the sixth Standard for Mathematical Practice.

In the area of measurement, students select appropriate tools, and they take care to use the tools accurately and effectively. Students need many opportunities to experience measurement tasks before they work with actual tools. The second-grade task in figure 6.2 asks students to understand the need for precision when measuring with a ruler and shows one student's response.

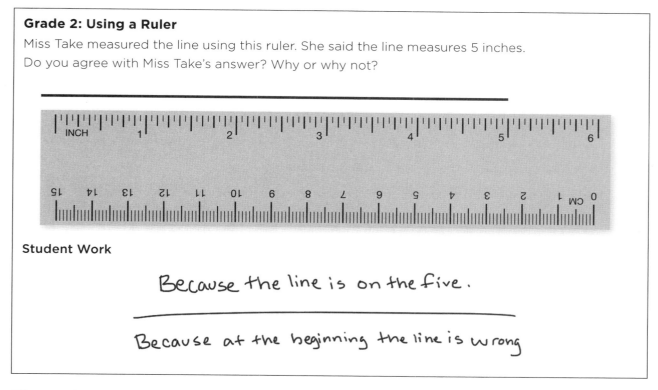

Figure 6.2: Using a ruler.

Visit go.solution-tree.com/MathematicsatWork for a free reproducible version of this figure.

The student in this example began answering the ruler task by saying the answer was correct. He looked again and drew a line and said, "Oh, no, that's not right . . ." Interestingly, he said that the line was wrong rather than the ruler being placed incorrectly. This student attended to precision as he reconsidered his first answer to correctly state that the measurement was wrong.

With calculations, students need to work accurately, use symbols appropriately, and make estimates as needed, depending on the context. Students also reflect on whether their approach to a particular task makes sense.

In figure 6.3 (page 150), students consider what the word *about* means in the task to determine which values are closest to 120. The number line can help students establish the position of the amount of money and then interpret the meaning of *about*. In this context, the application provides a connection to precise rounding.

Grade 3: What's in the Bank?

Amelia says that she has about 120 pennies in her piggy bank.

Which quantity of pennies might she possibly have in her bank?

(Circle the correct amounts. There may be more than one correct answer.)

111 pennies	118 pennies	109 pennies	100 pennies
116 pennies	210 pennies	128 pennies	122 pennies

Alicia said they could check their work by using a number line.

110   130

Explain how this number line can help determine the amount of pennies in the piggy bank.

Figure 6.3: What's in the Bank? activity.

Visit **go.solution-tree.com/MathematicsatWork** *for a free reproducible version of this figure.*

Two pieces of work for What's in the Bank? reveal students' thinking about this problem. In figure 6.4, the student finds values less than 120 and writes about using the number line to help find the values close to the desired number of pennies. In this case, the student does not consider amounts greater than 120.

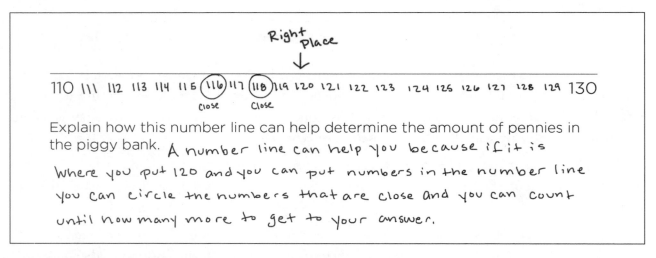

Figure 6.4: Sample student work 1 for What's in the Bank?

The student in figure 6.5 does not consider using the number line, and yet, this student does include values greater than and less than the stated value of 120, though not including the additional answer of 116. This student also states that he subtracted 2 in order to find the solution.

Each example of a conversation or task we show in this section provides an opportunity for students to use precision in the areas of communication, measurement, and calculation. Teachers can use the demonstrated errors in precision to plan next steps for students.

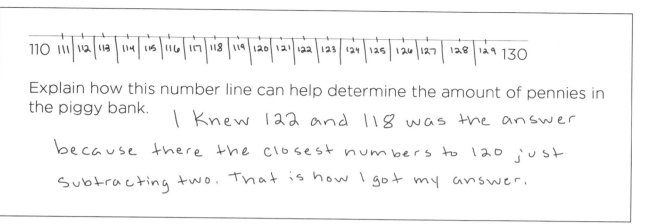

Explain how this number line can help determine the amount of pennies in the piggy bank.

I knew 122 and 118 was the answer because there the closest numbers to 120 just subtracting two. That is how I got my answer.

Figure 6.5: Sample student work 2 for What's in the Bank?

Table 6.1 shows more examples for what students do during a lesson to demonstrate evidence of learning Mathematical Practice 6 and what actions you can take to develop this critical thinking in students.

Table 6.1: Student Evidence and Teacher Actions for Mathematical Practice 6

	Student Evidence of Learning the Practice	Teacher Actions to Engage Students
Attend to precision.	Students: • Calculate accurately using varied procedures that make sense • Communicate effectively and with precision both orally and in writing • Use mathematics vocabulary appropriately • Write symbols and explain their meaning • Reflect and evaluate answers to ensure that they make sense • Attend to units and labels to clarify values and concepts • Select and use mathematics tools effectively • Explain when an exact answer is needed and when an approximation is sufficient	Teachers: • Promote mathematical language by using and modeling correct terminology • Encourage students to investigate errors and use these errors to deepen their understanding • Pose questions that challenge students to think about how precise a solution needs to be based on its context • Ask students to explain the meaning of the symbols used in their work • Prompt students to compare and explain their solution paths and answers • Review student work and provide feedback to students including next steps to take • Model the use of tools and provide tasks that require the use of mathematical tools

*Visit **go.solution-tree.com/MathematicsatWork** for a free reproducible version of this table.*

Understand *Why*

Precision is central to every lesson in mathematics. Consider the importance of it through the framework of communication, measurement, and calculation.

Communication

Attending to precision involves communication. Using the correct vocabulary in appropriate context enhances one's ability to communicate both orally and in writing. The English language, however, presents challenges. Consider the multiple ways that some common words such as *face*, *feet*, and *power* are used. In everyday life, *face* and *feet* can refer to our own body parts or those of an animal. Mathematically speaking, *face* refers to a polygon making up one flat surface of a three-dimensional shape, and *feet* represents a standard unit of measure. Think of the word *power*, and we generally think of strength or amount of electricity. *Power* in a mathematical context refers to an exponent. Homophones can also indicate different meanings. For example, *sum* and *some* can cause confusion.

The English language also presents challenges regarding the names of numbers. As students learn the names of numbers, the pattern does not become obvious early on in the sequence of numbers. In other languages such as Japanese, the counting sequence repeats at 10. *One ten one* represents eleven and *one ten two* is equivalent to 12. In English, students need to remember new words rather than relying on a pattern that makes sense. Eleven and twelve are confusing, and then students confront the *teen* numbers. Thirteen, fourteen, and so on can be problematic because students hear the unit and then it becomes attached to *teen*. For this reason, many young learners reverse the digits to represent 17. They hear *7* and then place the one. Learning the counting sequence requires students to attend to precision.

Just as language can be difficult, mathematical notation can also present inconsistencies. In *Embedded Formative Assessment*, Dylan Wiliam (2011) notes that the operation indicated by $8x$ is multiplication, but the operation indicated by $8\frac{1}{2}$ is addition. The subtle difference between these terms is not easily mastered. Students may connect $8\frac{1}{2}$ with addition by saying "8 and $\frac{1}{2}$." Notice that $8x$ is not read as "8 and x," it is read as "$8x$" or "8 times x." This term can also be stated as the "product of 8 and x." The connection of language with notation reinforces the meaning of each term.

Similarly, the meaning of the equal sign frequently confuses students. Some students working with the problem 4 + 3 = refer to the equal sign as an indicator that the answer should be next. In *Children's Arithmetic*, Herbert Ginsburg (1977) interviews young students about the meaning of the equal sign. Frequently, students state "equal sign means what it adds up to" and "equal sign means that you're coming to the end" (Ginsburg, 1977, p. 112). Of course, "what it adds up to" makes sense in terms of the addition, but it can lead to further challenges. Students who view the equal sign this way have difficulty when faced with the unknown value in varying positions in an equation, such as in the following.

$$\Box = 3 + 4$$
$$7 = \Box + 6$$
$$2 + 3 = \Box + 4$$

In the early grades, teachers should refer to the equal sign as "has the same value as" before the actual notation is introduced to help students understand its significance. Thus, precision with language and correct notation leads to increased understanding and ability to communicate effectively.

 Developing an expanding vocabulary requires multiple opportunities to experience the words in a variety of ways. Word and world knowledge are acquired incrementally, in bits and pieces here and there (Nagy & Scott, 2000). Nagy and Scott (2000) also suggest that teachers introduce vocabulary and word meanings in three ways: (1) photos or cartoon illustrations, (2) gestures or short dramatic scenarios, and (3) student-friendly definitions with deeper discussions of word meanings as needed.

Organizing vocabulary into a three-tier framework as articulated in appendix A of *Common Core State Standards for English Language Arts & Literacy in History/Social Studies, Science, and Technical Subjects* (NGA & CCSSO, n.d.) can benefit learners in terms of acquisition and retention. The first tier consists of everyday words such as *more, less, high, low, many, house, dog,* and so on. Tier two includes frequently occurring words in academic, cross-curricular contexts, such as *explain, explore, justify, compare,* and *predict.* Tier three vocabulary words are those low-frequency words that are specific to the subject area. In mathematics, words such as *function, denominator,* and *quotient* are in tier three. Kimberly Tyson (2013) suggests that "understanding the three tiers can help separate the 'should-know words (Tier 3)' from the 'must-knows (Tier 2)' and the 'already-known words (Tier 1).'" Checking in with students regarding tier one words and focusing on tier two words as they occur across disciplines can facilitate students' learning of vocabulary.

Students need many opportunities to share their understandings with their peers and use the vocabulary in the appropriate context. Margaret McKeown, Isabel Beck, Richard Omanson, and Martha Pople (1985) suggest that even as many as four "'teaching' encounters with a word do not give learners enough knowledge to improve their reading comprehension in text that contains the word. It may take as many as twelve experiences" (as cited in McEwan-Adkins, 2010, p. 139).

Research also reveals the importance of having students collaborate in small groups. (See the problem-solving strategy in chapter 1, page 13.) Providing time for students to share their understandings, to support each other in the development of their knowledge, presents surprising results. As Dylan Wiliam (2011) suggests, the effect of peer tutoring almost matches gains made when the teacher worked with just one student. John Schacter (2000) further details:

> A study of 109 students in fourth-, fifth-, and sixth-grade classrooms found that students working in student-led groups learned almost as much as students getting one-on-one tutorial instruction from a teacher, and those in student-led groups actually learned more than those in teacher-led groups. (p. 134)

Measurement

Students gain their understanding of measurement through a gradual process beginning with informal comparisons and continuing eventually to the use of tools and formulas. John Van de Walle and LouAnn Lovin (2006) offer three steps to help students develop a conceptual understanding of measurement.

1. Decide on an attribute to measure.

2. Select a unit with the attribute.

3. Compare the units, by filling, covering, matching, or some other method, with the attribute.

As students develop their understanding of measurement, the demand for precision increases. The use of informal comparisons such as more or less, shorter or longer, higher or lower, heavier or lighter, and so on lays the foundation for the eventual use of units. Measurement indicates the quantity of units that describes the specific attribute under investigation. As students develop their skills in using appropriate tools, it is important that they understand when to use the tool and how to use it. As Van de Walle and Lovin (2006) note, "In the sixth National Assessment of Educational Progress (Kenney & Kouba, 1997), only 24 percent of fourth-grade students could give the correct measurement of an object not aligned with the end of a ruler. . . . These results point to the difference between using a measuring device and understanding how it works" (p. 226).

Students need to consider the level of accuracy a particular context requires. When measuring the length of a book, what unit of measure will serve the purpose? How accurate should students be in measuring the height of the door or the length of one's foot? How will students measure area and volume?

It is important for teachers to include the approximate nature of measurement. By using words such as *about*, or *a little less than*, students can begin to realize that the actual measurement exists within a certain range. As John Van de Walle and LouAnn Lovin (2006) explain, "The use of approximate language is very useful for younger children because many measurements do not come out even. In fact all measurements include some error" (p. 228). For this reason, all measurements are truly approximations. As students continue to work with measurement, they realize that the smaller the unit of measure, the closer the measurement value approaches the actual value. They also understand that the smaller the units of measure, the greater number of units are needed.

As students work with measurement, they also need to focus on the appropriate use of labels. Recording the units associated with a solution, labeling axes, and writing a title for a graph all encompass the act of attending to precision.

Calculation

Students engage in Mathematical Practice 6 as they solve problems and use estimates. Students begin with whole numbers, and as they progress through the grades, they use decimals and fractions. As students develop these skills, they focus on the precision of the answer that is needed. Students determine this precision as they work within a particular context. For example, consider the following problem.

> James wants to buy 6 items each costing $3.25. He has $20 in his wallet.
> Does he have enough money to buy all 6 items?

A student does not necessarily have to compute 3.25 × 6 exactly and can, instead, start with 3 × 6 = 18. The context of money helps students make sense of the mathematics. Now they can think about the value of 6 quarters rather than 0.25 × 6 and know $18.00 + $1.50 is less than $20 so James has enough money.

Estimation presents opportunities for students to judge the accuracy of their answer if they have approximated the answer first. A student estimates that the answer to 10 × 0.25 must be closer to 2 than 5. The student reasons that 0.25 is the same as $\frac{1}{4}$ and 10 × $\frac{1}{4}$ must be less than 5 because $\frac{1}{2}$ of 10 is 5. The student might continue to acknowledge that $\frac{1}{4}$ *is* $\frac{1}{2}$ of $\frac{1}{2}$ and that half of 5 is closer to 2 than 5.

Fluency is another factor when considering students' calculation skills. Procedural fluency is defined by the Common Core State Standards for Mathematics as "skill in carrying out procedures flexibly, accurately, efficiently, and appropriately" (NGA & CCSSO, 2010, p. 6). Similarly, Arthur Baroody (2006) states the fact that fluency is "the efficient, appropriate, and flexible application of single-digit calculation skills and . . . an essential aspect of mathematical proficiency" (p. 22).

Attending to precision involves clear communication using the appropriate vocabulary, working with measurement and estimation, and making accurate calculations. It is also important that students determine whether their solutions make sense.

Strategies for *How*

As noted, attending to precision focuses on three main areas: communication, measurement, and calculation. Both measurement and calculation involve estimation. The following strategies provide teaching suggestions and student activities to engage the learners in this sixth Standard for Mathematical Practice.

 ## *Precise Teacher Language*

As stated, teachers need to model the precise use of language. For example, avoid using the term *top number* when referring to the numerator of a fraction. Using *top number* can lead students to the misconception that fractions are made up of two whole numbers. Students should experience the correct mathematical terms even in their early years. Young children are so receptive to language; teachers need to always demonstrate the use of correct mathematical language.

Robert Marzano (2006) reports that direct instruction that is focused on vocabulary development can improve students' ability to read and understand specific content. He offers the following six-step process for teaching academic vocabulary:

1. Provide a description, explanation, or example of the new term. (Include a non-linguistic representation of the term for ESL kids.)

2. Ask students to restate the description, explanation, or example in their own words. (Allow students whose primary existing knowledge base is still in their native language to write in it.)

3. Ask students to construct a picture, symbol, or graphic representing the word.

4. Engage students periodically in activities that help them add to their knowledge of the terms in their notebooks.

5. Periodically ask students to discuss the terms with one another. (Allow in native language when appropriate.)

6. Involve students periodically in games that allow them to play with terms. (as cited in Newaygo County Regional Educational Service Agency, n.d.)

Teachers also need to be careful about using expressions that occur out of convenience rather than a mathematical convention. Consider the phrase *goes into*. There is a possible confusion when a teacher asks "8 goes into 16 how many times?" What operation is "goes into"? If a student is thinking about division, "how many times" might cause that same student to use multiplication. Instead, consider asking "How many sets of 8 are needed to make 16?" or "How many groups of 8 are needed to make 16?" Using this wording will help students later understand $4 \div \frac{1}{2}$ as "How many sets of $\frac{1}{2}$ are needed to make 8?" so they understand the quotient is 8. Subtle changes in the wording allow students to build their conceptual understanding of division and its relationship with multiplication.

Another example involves decimal fractions. It is important to avoid using the word *point* to indicate the decimal. When reading 12.04, be careful to say "12 and 4 hundredths," which is a match to the decimal fraction $12\frac{4}{100}$. The decimal point is referenced by the word *and*. For this reason, try to avoid using *and* when reading large whole numbers. For example, 2,322,062 should be read as "two million, three hundred twenty-two thousand, sixty-two." Too often the word *and* is slipped in as the number is read. Attending to precision reinforces students' place-value understanding. Consider also avoiding terms such as *timesed* and *minused*. (See appendix H, page 265, for other terms lacking precision.)

Same Words—Different Meanings

There are many mathematical terms that have different meanings outside of mathematics. For example, *left*, *round*, and *base* all have mathematical definitions and are also used in everyday language, as table 6.2 shows. Classes can keep a running list of such terms along with their varying uses in our daily lives.

Table 6.2: Vocabulary Used in Mathematics and in Other Contexts

Term	Mathematics Example	Context Examples
Left	Remainder: When you divide 5 by 3, how many are left? Difference: How many are left?	Directions: Turn left. Location: My wallet is in my left pocket.
Round	Approximation: Round to the nearest tenth. Geometry description: How do you describe the edge of a circle?	Sports: Rounds in the playoffs; number of golf games played as in "How many rounds did you play this season?"
Base	Exponential use: 3^2 3 is the base. Geometry: Base of a triangular prism	Sports: Base as in baseball

Sometimes tasks have the same words but are solved using different operations. Consider the following examples.

- 30 is half of what number?
- Half of 30 is what number?

Students need to focus on the meaning of each example. For each pair of phrases, students need to make sense of what is asked. Students can predict the size of the answer, and this can help them make sense of the context. Students also attend to precision as they make sense of each situation and determine if their calculation is accurate. Careful reading to interpret the critical phrases will enable students to be successful on this exercise. Should they double 30 or divide it by 2?

Elementary teachers have stressed finding key words as a problem-solving strategy. As discussed in the Mathematical Practice 1 strategy Reading Word Problems, emphasizing key words can be problematic (see page 21). The examples in figure 6.6 further illustrate the danger of relying only on key words.

Grades 3–5 Task: Making Sense

The same words can have the same meanings but may lead to different solutions.

For each problem, estimate whether the answer will be greater or less than the underlined value. Then determine the solution.

Question	Greater or Less Than Underlined Value	Solution
1. A. Half of <u>40</u> is what number? B. <u>40</u> is half of what number?		
2. A. How many more is <u>8</u> than 3? B. Which number is <u>8</u> more than 3?		
3. A. <u>7</u> is 16 less than what number? B. <u>7</u> is how many less than 16?		
4. A. <u>23</u> decreased by 9 is what number? B. <u>23</u> decreased by what number is 9?		
5. A. <u>12</u> increased by 18 is what number? B. <u>12</u> increased by what number is 18?		

Select one of the sets of questions from 1–5. Explain your reasoning as to how you determined the answers for each question.

Figure 6.6: Making sense example.

*Visit **go.solution-tree.com/MathematicsatWork** for a free reproducible version of this figure.*

Students need to attend to the precision of the language in order to make sense of each question. The phrases are similar; what is known and what is unknown differ in each group.

Homophones

Homophones represent another source of confusion. Some terms in mathematics sound exactly like other words that have different meanings and are spelled differently. It is helpful for students to review these words in a context, as figure 6.7 illustrates.

Context to the Rescue!

Select the appropriate word to place in each sentence.

1. Select: *sum* or *some*

    Sarah showed her brother _____ of her magic tricks.

    Clarence found the _____ of all the money in his pocket.

2. Select: *whole* or *hole*

    Marco reasoned that 4 pieces represented one-half so there must be 8 pieces in the _____.

    The dog dug a _____ in the ground so that he could hide his bone.

3. Select: *our* or *hour*

    Melinda said, "It is such fun when _____ family gets together."

    Waiting at the doctor's office, a patient said, "I am so tired of waiting. I have already been here for one _____."

4. Select: *pair / pear / pare*

    Mohammed asked his brother to _____ the potatoes before cooking them.

    Michael grabbed a _____ of socks that he had in his drawer.

    Benjamin chose to eat a _____ instead of an apple.

Figure 6.7: Context to the Rescue! example.

*Visit **go.solution-tree.com/MathematicsatWork** for a free reproducible version of this figure.*

The most common homophones in elementary classrooms are *sum* and *some*. The learning progressions for the Common Core mathematics suggest that the use of *sum* be delayed until first grade (Common Core Standards Writing Team, 2013) because kindergarten students think of *some*, as in "some pretzels from the bag" or "some students are in line"—a part rather than the whole or total—when hearing the word as opposed to reading it. Teachers can clarify the distinction by using the term *total* to indicate the sum in kindergarten.

Measurement Activities

Measurement activities also require the use of precise language. As students measure objects with non-standard and standard units, it is important to identify the units. Students attend to precision when they recognize the difference between square inches and inches as well as inches and feet. For example, How many square inches are in one square foot?

Students also attend to precision as they select appropriate tools. Given a choice of using a string, raw spaghetti, index card, scissors, pencil, or paper clips, which tool might a student select in order to determine the area of the cover of their science textbook? As the cover is two-dimensional, a tool having those same attributes would be appropriate.

Precise Student Language

Not only is precise teacher language needed in the classroom, but precise student language is essential as well. The following sections discuss strategies for developing precise student language.

Back-to-Back

Students see the power of using precise language as they give directions. Back-to-Back requires that students sit with their backs toward one other and give directions or explain solutions or work. For example, one student might use geometric shapes to create a design. This student then describes the design to his or her partner. The partner listens and tries to duplicate the original design based on the first student's directions. After the last direction is completed, students compare designs and receive the instant feedback on how well the directions were given and followed. Teachers can monitor the activity by listening for key vocabulary words such as *trapezoid, perpendicular, parallel, rotate,* and *adjacent.* Teachers can then share these words with the class as a way to summarize the work and emphasize the precision in their language. Students can also use Back-to-Back to practice orally communicating their solution and strategy on a problem before sharing it with the whole class.

Reciprocal Teaching

When students are challenged to teach others, both partners benefit. As John Hattie (2009) suggests in *Visible Learning,* reciprocal teaching is one of the most powerful ways to deepen students' understanding. As students begin to realize that their partner is relying on them for information and support, they feel more accountable. Formats for reciprocal teaching vary; Time to Teach and Write It Right are just two examples.

Time to Teach

Groups of four students collaborate to solve word problems. Each student is given one of four types (addition with the result unknown or change unknown or subtraction with the result unknown or change unknown) to become the expert for the group. Figure 6.8 (page 160) shows a grade 2 example from the CCSS for mathematics appendix.

	Result Unknown	Change Unknown
Add to	Two bunnies sat on the grass. Three more bunnies hopped over there. How many bunnies are on the grass now? $2 + 3 = ?$	Two bunnies were sitting on the grass. Some more bunnies hopped over there. Then there were five bunnies. How many bunnies hopped over to the first two? $2 + ? = 5$
Take from	Five apples were on the table. I ate two apples. How many apples are on the table now? $5 - 2 = ?$	Five apples were on the table. I ate some apples. Then there were three apples. How many apples did I eat? $5 - ? = 3$

Source: NGA & CCSSO, 2010, p. 88.

Figure 6.8: Word problem types.

Each student in the group is responsible for solving one specific type of problem. Teachers provide several examples of each type of problem. All students who are working on "Take from and Change unknown" problems work together to do the problems and check each other's answers. All students working on each type of problem are accountable to each other. All group members must be able to solve the problem as well as explain their thinking. After practicing, students then go to their original groups and begin teaching their teams about their problem types. Once all team members have shared their strategies, the team practices each type of problem together. Students are then given problems in which the four types are mixed up. Students work independently at first, and then they compare their answers with their team members'. The team then decides which type of problem they need to continue to practice. The following list summarizes the five steps for Time to Teach.

1. Students are given one problem type; each student in the group has a different problem type.

2. Those students with the same problem type work together on problems, ensuring that they all understand the problem type and can explain this problem to their original groups.

3. Students return to their original groups. They support their groups working through their problem type. Each expert may teach the team as needed.

4. Teams are then given practice problems in which the different types are mixed up on the paper. Students work independently at first and then compare their answers.

5. Teams determine which problems that they need to continue to practice.

Write It Right

Write It Right is a partner activity in which one student tells his or her partner how to solve a particular problem. That first student tells the second student what to record, how to record it, and the reasoning for doing so. With each new problem, the partners change roles. After completing a set of problems, students can then exchange their work with another pair of students. This group of four compares their solution strategies and justifies their reasoning. Throughout this activity, students focus on using correct language, giving clear directions, and making certain that their answers make sense.

Symbol Literacy

Both students and teachers need to pay close attention to the use of symbols. Consider the inequality signs used in elementary classrooms to compare values. Many teachers hope to help students distinguish between these signs by referring to an alligator, as in "The alligator eats the bigger number." Although attention grabbing, the alligator is not a mathematical tool. Instead, teachers can simply say that the sign points to the smaller number.

Another symbol that is difficult for students to use and read correctly is the division sign. Compare (a) $12 \div 3$ and (b) $3\overline{)12}$. Both are read as "twelve divided by three." Thus, (a) is read from left to right while (b) is read right to left. As a result, students and teachers tend to use the phrase "goes into" for example (b). Along with 12 divided by 3, there are several other correct ways to interpret the division including:

- 3 divided into 12
- Take $\frac{1}{3}$ of 12
- $\frac{12}{3}$

The three preceding examples provide students with opportunities to make the connection between division and multiplication, and students may begin to develop an intuitive understanding of the meaning of *reciprocal*. It is also helpful to present the fractional representation using a horizontal bar rather than a slanted one. 12/3 is easily typed, but the connection between the fractional representation and division is not as easily recognized as compared to $\frac{12}{3}$. It is also important to use the horizontal bar so later on $\frac{1}{x+1}$ is not interpreted as $\frac{1}{x} + 1$, which is likely if the expression is written $1/x + 1$.

The equal sign is another symbol that needs to be carefully developed. As mentioned earlier in this chapter, students tend to think of the equal sign (=) to indicate that the answer comes next, as in $7 + 8 = \underline{\hspace{1cm}}$. You can model the understanding of the equal sign as "is the same value as" so that students are successful with statements such as $7 + 8 = \underline{\hspace{1cm}} + 3$. The Does It Fit? activity for students in grades 1 and 2 in figure 6.9 (page 162) provides opportunities for students to develop an understanding of the equal sign as a balancing point between the expression on each side of the equation.

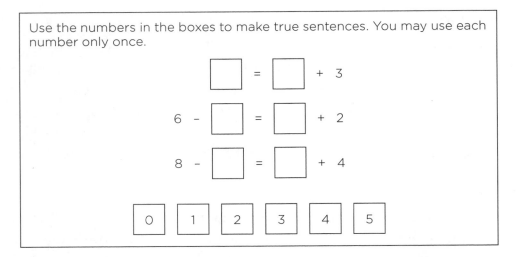

Use the numbers in the boxes to make true sentences. You may use each number only once.

☐ = ☐ + 3

6 – ☐ = ☐ + 2

8 – ☐ = ☐ + 4

| 0 | 1 | 2 | 3 | 4 | 5 |

Figure 6.9: Does It Fit? example.

*Visit **go.solution-tree.com/MathematicsatWork** for a free reproducible version of this figure.*

When first working with fractions, it is important to delay the use of the symbolic representation. When students first partition shapes into the equal sized groups, teachers should refer to fourths, thirds, or halves as compared to $\frac{1}{4}$, $\frac{1}{3}$, and $\frac{1}{2}$. For example, if the symbolic representation of one-fourth is introduced before students have developed an intuitive understanding of the meaning, then they are likely to think that $\frac{1}{4}$ is greater than $\frac{1}{2}$, as students are prone to apply their understanding of whole numbers and 4 is greater than 2 (Small, 2014).

Mathematics Vocabulary

Vocabulary development provides opportunities for increased understanding of the mathematics and increases students' ability to communicate their ideas. Using the appropriate language reflects another aspect of Mathematical Practice 6. Consider the following three strategies to engage elementary students as they build their vocabulary in mathematics: (1) using vocabulary organizers, (2) creating active word walls, and (3) playing games.

Using Vocabulary Organizers

Chapters 1 and 2 provided specific graphic organizers related to problem solving and reasoning. The organizers for precision help students focus on vocabulary. Figure 6.10 shows a graphic organizer to use with mathematics vocabulary.

For another version, consider replacing "Example" and "Nonexample" with "Meaning" and "Sentence." Students can express the meaning of the vocabulary word using their own thoughts and then use the word appropriately in a sentence. An additional graphic organizer in figure 6.11 provides students with opportunities to connect with the word on a personal level as well as reflect on the context in which the word is typically used.

Word	Picture
Example	Nonexample

Figure 6.10: Vocabulary graphic organizer.

*Visit **go.solution-tree.com/MathematicsatWork** for a free reproducible version of this figure.*

Word	Context Clue	Definition	Picture	Connection to Me

Figure 6.11: Making Connections organizer.

*Visit **go.solution-tree.com/MathematicsatWork** for a free reproducible version of this figure.*

Creating Active Word Walls

Word walls display vocabulary words in a classroom. In order to support student learning, make word walls accessible to students and use them actively. Teachers should organize the word wall according to the current unit of study. As a new vocabulary word is introduced, teachers can ask students to locate the word on the wall. Students then discuss what they think the word might mean in the context. Students need to easily read and access the words, so they should not be placed too high on the wall. Additionally, be cautious with the number of words on the word wall. When too many words are on the wall, students find it difficult to focus on a single word or group of words.

Figure 6.12 (page 164) provides an example of a word wall used during an introductory lesson to fractions. Note the multiple representations of the unit fractions.

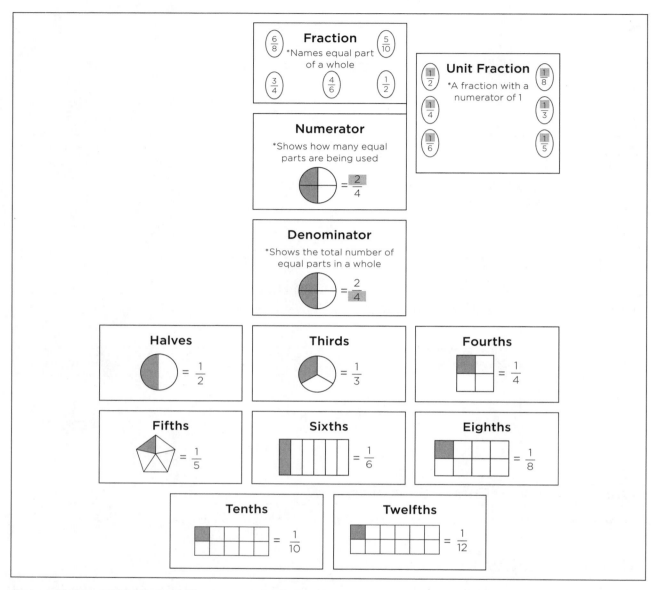

Source: Tammy Farris, E. Pole Elementary School, Taunton, MA. Used with permission.

Figure 6.12: Active word walls.

Teachers can implement a signal to celebrate every time a student uses the vocabulary from the new unit. For example, students may applaud and clap their hands in a circle—giving a "round of applause." Some teachers celebrate by playing chimes or ringing a small bell. As a student uses the vocabulary word appropriately, ask other students to restate what the first student said. Students might also take turns drawing a picture to show the meaning of the word on the wall. When students begin to develop an understanding of the vocabulary words for the unit, they can use one of the graphic organizers to record their current understanding of the word.

Word walls provide a source for students to reference as they sort and group vocabulary words. Teachers can even add a few review words along with the current vocabulary. Pairs of students record the words that they have grouped and why they think the words belong together. Students then share their thoughts with other pairs of students.

Word walls can also engage students in classroom activities. For example, in a geometry unit, the focus words might be *trapezoid, rhombus, quadrilateral, square, triangle, polygon, parallelogram*, and *circle*. Students consider how to order these words from the most general category to the most specific. Teams work together to frame their thoughts. Then, one team moves one word on the wall to a new location. Without any discussion, another team moves a word. This process continues until the class is satisfied that it has grouped the vocabulary according to the specific guidelines. If teams disagree by moving the same word back and forth between places, the teams then present their reasoning to the class. (See figure 6.17, pages 169–172, as this activity is included in the closure of the lesson.) Finally, students can serve as an active word wall themselves. Students wear one vocabulary word like a sign. Teachers can use name tags or make signs from construction paper and string, and students can draw pictures to represent the word on their signs. Throughout different parts of the day, the teacher asks students about their understandings of their words, and students explain their pictures or use the vocabulary words in a sentence.

Playing Games

Literacy games are also useful in mathematics, including the following three: (1) Concentration, (2) HORSE, and (3) Make a Sketch.

Concentration involves cards with a word on one side and a matching picture or sentence using the word on the other side. Two students place cards facedown and then take turns trying to find two cards that match. After each try, the student turns the card over and puts it in its original place—either faceup with a match or facedown when they don't match.

HORSE is another vocabulary game for two people. One student selects a word from the word wall or list. This student provides blank spaces for each letter in the word. The second student guesses a letter, and if it is in the word, the first student places the letter in the appropriate space. If the second student's selected letter is not in the word, the student receives the first letter in the word *horse*. The game continues until either the second student guesses the vocabulary word correctly or all of the letters in *horse* have been spelled out.

Make a Sketch is a game that requires students to work in small teams. One person from each team makes a sketch or drawing of the teacher-selected vocabulary word or expression. Each team gets a point when a team member correctly determines the vocabulary word based on the drawing. Sketchers may not use letters as part of their drawings. They may use sound-alike words and indicate to their team that they are doing so by touching their ear. Each student has a chance to be the sketcher within the game. When all students have had the opportunity to draw, the game is over. The team with the most points wins. Consider using the following words when playing Make a Sketch.

- *sum*
- *total*
- *multiplication*
- *graph*
- *difference*
- *product*
- *fraction*
- *equation*
- *divisor*
- *plot*
- *numerator*
- *rectangle*

- *circle*
- *cube*
- *triangle*
- *vertex*

- *face*
- *trapezoid*
- *tens*
- *tenths*

- *million*
- *decimal point*
- *place value*
- *exponent*

- *base*
- *estimation*
- *calculator*

The time teachers dedicate to helping students develop their understanding and appropriate use of vocabulary contributes to their ability to attend to precision. The appropriate use of vocabulary enables students to communicate effectively and deepen their understanding of the concepts involved.

Estimation as a Tool

Estimation, to be effective, needs to serve a purpose. Instead of estimating and then calculating, many students simply calculate and then round their answer to use as an estimate. In these instances, students may be trying to avoid a mistake or they may not understand the purpose of making an estimate. By placing estimation in a context, it then becomes a tool to help answer specific questions, such as the following.

- Do you have enough money to purchase a model airplane and the book?
- Did you buy enough beads to make a bracelet for your sister?
- Will we arrive on time?
- How much paint do we need to cover this wall?

In each question, students attend to precision as they consider their need for accuracy. Does their answer require an exact amount? Can they be close? How close? When purchasing the airplane and book, students can round prices higher to the nearest dollar and then compare the total with the amount of money in their wallet. How does knowing the number of square feet of the wall help determine the number of pints or gallons of paint they need? Such open questions stimulate students' thinking and focus their attention on the degree of precision needed to answer the given question.

Estimation can also serve as a check (see the strategy for Mathematical Practice 1, Estimating Upfront, page 30). Students learning to work with decimals can use estimation to determine whether their answer makes sense. Students make a quick estimate. They then perform the calculations. Students compare their calculated results with their approximations. Figure 6.13 shows a chart to help students estimate as a check.

	Estimate the answer. Explain your thinking.	Calculate the answer. Is your answer close to your estimate?
1. $12.4 \times 10.08 = \square$	My answer should be about _____ because _____.	
2. $\square = 23.6 \div 0.2$	My answer should be about _____ because _____.	

Figure 6.13: Estimation as a check—grades 3–5 task.

Error Analysis

Students will engage in learning when asked to find an error. Teachers can create the error themselves or use actual students' work to share with the class. As mentioned earlier in Mathematical Practice 3, the ideal classroom climate contributes to students viewing errors as an opportunity for learning. With an understanding that all learners become better mathematicians as they recognize and correct their mistakes, students can readily receive feedback. Building such a community of learners takes time. It may be necessary to change names or cross out the student's identity as this activity is first introduced. When students identify the mistake, they also need to offer a suggestion to avoid the error in the future. See figure 6.14 for examples.

I counted the ducks in the pond when I arrived. In a little while, 12 more ducks flew into the pond. Now, there are 21 ducks. How many ducks were in the pond when I arrived?

Sally's Work

$$\begin{array}{r} 21 \\ -\ 12 \\ \hline 11 \end{array} \qquad 11 \text{ ducks}$$

What mistake did Sally make?   Show Sally a different way to solve this problem.

Michael's Work

"I counted up—12, 13, 14, 15, 16, 17, 18, 19, 20, 21.
There were 10 ducks in the pond."

What mistake did Michael make?   Show Michael another way to do this problem that is different than the method you shared with Sally.

Figure 6.14: Find the Error examples—grades 1 and 2.

Students can work with partners to discover the errors and make suggestions as to the different ways to solve the problem. Partners can then share their work with another group to further extend their opportunities to explain their thinking. Students are attending to precision as they identify the error and explain their thinking.

Clickers are another error-analysis tool that can help students and teachers identify mistakes. Students respond to a question by selecting one of the multiple-choice answers, and the class's results are displayed. Students identify the most frequently selected wrong answer and determine what mistake a student made who chose that answer. Following is an example of a fifth-grade question and multiple-choice answers.

Calculate: $18 - 6 \div 3 + 1$

    a. 5

    b. 17

    c. 9

    d. 3

Additionally, students can be asked to insert parentheses so that they can obtain one of the other options presented in the multiple-choice selections. For example, if asked to insert parentheses so the answer is (d) 3, students can write the following.

$$(18 - 6) \div (3 + 1)$$

Calculator Feedback

Calculators offer vast opportunities for students to explore, look for patterns, and deepen their understanding of numerical relationships. Teachers give students a situation and ask them to make a prediction. The calculator then gives them feedback on the accuracy of their predictions.

Students can also use calculators to check their answers to problems. Teachers can create a calculator station for students to go to and check their work. By requiring students to physically move to the station, they are not tempted to use the calculator as they are finding the answers initially. Structured movement within a class period provides additional benefits such as a quick break and a change of pace.

Consider using the following two calculator activities with students: (1) Target Practice and (2) Make It Happen.

Target Practice

Teachers present students with a starting value and the target value. As figure 6.15 illustrates, they are to use multiplication and one factor to reach their target number. Partners take turns; they must begin by using the value from the previous calculation as a starting point. The goal is to reach the target number within a range of 1.5 above or below the target.

Game 1	Starting Number	Multiplication	Target Number is 85.6
Student 1	41.6	41.6 × 2	83.2
Student 2	83.2	83.2 × 0.2	16.64
Student 1	16.64	16.64 × 5.1	84.64

Figure 6.15: Target Practice example.

*Visit **go.solution-tree.com/MathematicsatWork** for a free reproducible version of this figure.*

After the first student multiplied by 2 and found a product that was reasonably close to the target number, the second student thought he only needed a little bit more, and therefore, chose to multiply by 0.2. They were not alone in their sudden realization that multiplying by such a small value actually decreased the resulting product. Throughout this activity, students attended to precision of the factors and resulting product, and they shared their conclusions at the end of the activity.

Teachers can use Target Practice with any operation. After developing greater proficiency with multiplication of decimals, this group of students can proceed to division with decimals. It would be interesting to hear what they predict the patterns with division might be, knowing what they have discovered in terms of multiplication.

Make It Happen

Students have cards showing values such as 7; 70; 700; 7,000; 70,000; 700,000; and 7,000,000. Working in pairs, each student takes one card. Each student then places his or her value in a calculator. Using one operation, each student needs to make the calculator display show the value of his or her partner's card. Each student creates a chart like the one in figure 6.16 and records his or her work in the appropriate column. Then, students compare their results. Points are awarded when students correctly create the value on their partner's card. In this case of figure 6.16, both partners are awarded one point.

Partner A's card	Partner A's work	Partner B's card	Partner B's work
7	7 × 1000	7000	7000 ÷ 1000

Figure 6.16: Make It Happen—grades 3-4 example.

Lesson Example for Mathematical Practice 6: Attending to Attributes

The lesson plan in figure 6.17 focuses on attending to precision. Grade 4 students will demonstrate an understanding on the attributes of geometric shapes, specifically squares, rectangles, parallelograms, rhombi, and trapezoids. Students review the names of the shapes and build their understanding of the term with its attributes. Students then extend their understanding to identify common, measureable attributes of quadrilaterals and sort shapes based on those attributes.

Although Mathematical Practice 6 is the focus of the lesson as students sort the shapes and look carefully to identify the common attributes, students will also build an argument to justify their thinking with their partners (Mathematical Practice 3). Students work independently at first on each of the three tasks and then share their thinking with their partners. A commentary follows the lesson providing more information related to the rationale and importance of each lesson component. Figures 6.18–6.20 (pages 173–175) support the lesson's tasks.

Unit: Understanding categories of shapes (4.G.2)
Date: December 10
Lesson: Attending to Attributes

Learning objective: As a result of class today, students will be able to recognize attributes of shapes and sort the shapes into groups using their attributes.

---

continued →

Essential Standard for Mathematical Practice: As a result of class today, students will be able to demonstrate greater proficiency in which Standard for Mathematical Practice?

Mathematical Practice 6: "Attend to precision."

- Understand the meaning of *attribute*.
- Focus on the attributes of shapes.
- Compare and sort shapes based on their attributes.
- Use appropriate vocabulary.

Formative assessment process: How will students be expected to demonstrate mastery of the learning objective during in-class checks for understanding teacher feedback, and student action on that feedback?

- Students will review vocabulary of two-dimensional polygons, focusing on quadrilaterals.
- Students will identify shapes by name and will demonstrate their understanding that quadrilaterals can be grouped into categories.

Probing Questions for Differentiation on Mathematical Tasks

Assessing Questions	Advancing Questions
(Create questions to scaffold instruction for students who are stuck during the lesson or the lesson tasks.)	(Create questions to further learning for students who are ready to advance beyond the learning standard.)
• What is one attribute of this square?	• How is a rhombus like a square?
• What attributes do these shapes have in common?	• How is a rhombus different than a square?
• Select a shape that has two sets of parallel lines.	• How is a rhombus like a parallelogram? How is it different?
• How are these two shapes different?	• Is a square also a rectangle? What attribute does a square have that a rectangle does not have to have?
• What attributes do parallelograms have?	

Tasks (Tasks can vary from lesson to lesson.)	What Will the Teacher Be Doing? (How will the teacher present and then monitor student response to the task?)	What Will Students Be Doing? (How will students be actively engaged in each part of the lesson?)
Beginning-of-Class Routines How does the warm-up activity connect to students' prior knowledge, or how is it based on analysis of homework?	Teacher will review the basic names of shapes by giving each student geometric shapes, including quadrilaterals, triangles, and circles. Ask students to pick up the shape that you request. Ask for square, rhombus, triangle, rectangle, circle, parallelogram, and trapezoid. (continued)	Students arrange the shapes in front of them. As the teacher mentions a shape, the student picks up that shape. Students compare shapes with their partners. Do they all select the same shape when asked for a parallelogram?

	Ask students to compare the shape that they selected with the shape that their elbow partner selected. Are they the same or are they different?	
	Now ask students to select the shape that has four right angles. Do they all select the same shape? Ask students to select the shape that has four equal sides. Discuss with students that the rhombus and square both have four equal sides. They share this attribute. Ask students to hold up a quadrilateral whose attribute includes one pair of parallel sides.	Students select shape with four right angles. Students compare the shapes that they selected. Turn and talk with partner: What does *attribute* mean?
Task 1 How will students be engaged in understanding the learning objective? (See figure 6.18.)	Have students sit with a partner, and hand out copies of task 1 (figure 6.18). Share the learning target with students. Tell them that they will work independently at first and then share with a partner. Teacher monitors students' work by roaming the room and listening unobtrusively to students' conversations. Look for students who shade the rectangle and circle it as the rectangle is in the category of parallelograms.	Students work independently on task 1. Students identify the parallelograms by lightly shading them. They circle the rectangles and place a star on the trapezoids. Students check each other's work to ensure that the quadrilaterals are labeled correctly.
Task 2 How will the task develop student sense making and reasoning? (See figure 6.19.)	Ask students to work independently at first and then compare their work with a partner.	Students follow the clues in order. They may use pattern blocks and remove a block that no longer fits as a result of the next clue. Students write an explanation of why they know that their answer is correct. They compare results with a partner. If time permits, they create their own set of clues and exchange their puzzles with classmates.

continued →

Task 3 How will the task require student conjectures and communication? (See figure 6.20.)	Have students refer to Recognizing Shapes (task 1, figure 6.18) used earlier in the lesson, and remind students of the meaning of the intersection of the two circles in the Venn diagram.	Students write the label of the polygon in the appropriate section of the Venn diagram based on the headings above each circle.
Closure How will student questions and reflections be elicited in the summary of the lesson? How will students' understanding of the learning objective be determined?	Review the learning target. Have students reflect on the learning target using the Fist to Five method, where a fist indicates no understanding and one, two, three, four, or five fingers extended indicate the level at which they feel they totally understand and can help others. Use the word wall, and ask students to sort the quadrilaterals into groups. Have students work with their partners to organize their thinking. Encourage them to sketch a diagram of how the quadrilaterals relate to each other. Facilitate a discussion about teams' choices regarding the placement of the words on the wall. Give students a copy of the first activity. Ask them to critique their own work. Are there changes that they now want to make?	Students write a response to the following problem. • A rhombus and a square share many attributes. List as many as you can. • Which attribute do a rhombus and a square not have in common? One student from each pair silently goes to the wall and places a word in a new space. One partner from the next team places a word in a new space or near the first word if this student thinks that the two shapes are related. Each team arranges the words silently until it appears that the class has reached an agreement or one word is moved several times, thus indicating disagreement. Students review their work from the beginning of the lesson. What changes do they now want to make? Why?

Source: Template adapted from Kanold, 2012c. Used with permission.

Figure 6.17: Grade 4 lesson-planning tool for Mathematical Practice 6.

*Visit **go.solution-tree.com/MathematicsatWork** for a free reproducible version of this figure.*

The first task demonstrates whether students identify shapes using their appropriate names as well as recognize that shapes belong to specific categories, such as a rectangle is a special parallelogram. This opening activity gauges students' current understanding of the categories of quadrilaterals and can be used as a reflection at the end of the lesson.

Task 1: Recognizing Shapes

- Lightly shade the parallelograms.
- Circle the rectangles.
- Place a star on the trapezoids.

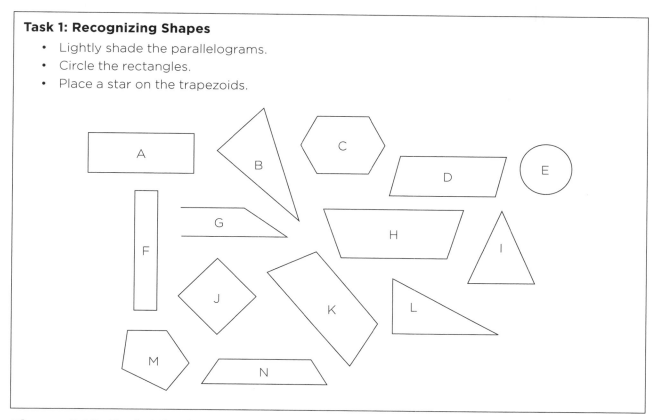

Figure 6.18: Task 1 for lesson Mathematical Practice 6.

*Visit **go.solution-tree.com/MathematicsatWork** for a free reproducible version of this figure.*

In the second task, students identify specific shapes based on clues presented in "Who am I?" Partners collaborate to select all of the quadrilaterals that match the opening clue and then weed out the shapes that no longer have the attributes mentioned in each subsequent clue. It is helpful to have pattern blocks as a reference for students.

In the third task, students sort the same shapes seen in the first exercise in the lesson. In this task, students place the shape's label in the appropriate section of the Venn diagram according to the indicated attribute listed in the title of each circle. Students need to pay attention to the intersection of the two circles, as they can only place those shapes with both identified attributes there.

Teachers have options for the lesson closure. Regardless of what options they choose, they should emphasize Mathematical Practice 6, "Attend to precision." They can present task 1 to see how students' thinking has changed as a result of the lesson. They can also use the word wall to regroup the quadrilaterals into categories that students identify. Vocabulary to have on the word wall includes: *quadrilateral, trapezoid, parallelogram, kite, rectangle, rhombus,* and *square.* Consider including pictures similar to the one in figure 6.21 as a reference for students. Be certain to remind students that the tick marks on the shapes indicate congruency, and a small square in an angle identifies a right angle.

Task 2: Who Am I?

Read the clues one at a time. Identify the shape the clues describe. Write a statement at the end to explain your answer.

1. I am a quadrilateral.

   I have at least one right angle.

   Both of my opposite sides are parallel.

   All my sides are the same length.

   > Who am I? This shape is a _____.

   > Explain your answer:

2. I have four sides.

   I have one pair of parallel sides.

   I have one pair of opposite sides that are equal in length.

   > Who am I? This shape is a _____.

   > Explain your answer:

3. I am a quadrilateral.

   My opposite sides are parallel.

   My opposite sides are the same length.

   I do not have a right angle.

   > Who am I? This shape is a _____.

   > Explain your answer:

4. Create your own clues to describe a shape. Share your clues with a partner. Do you both get the same answer using your clues?

Figure 6.19: Task 2 for Mathematical Practice 6.

Visit **go.solution-tree.com/MathematicsatWork** *for a free reproducible version of this figure.*

Task 3: Sorting Shapes

Look at the shapes used in task 1, Recognizing Shapes. Place the letter identifying the shape in the appropriate section of the Venn diagram. You may not be using all of the shapes in each question.

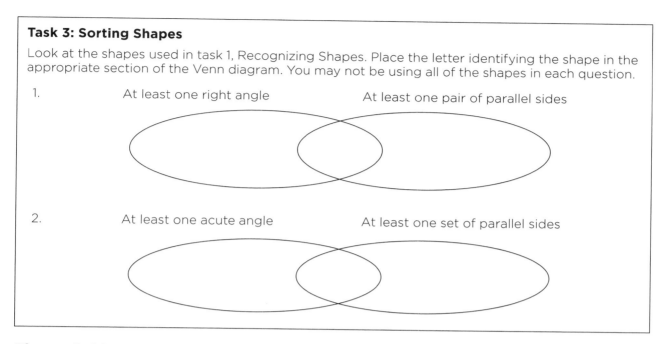

Figure 6.20: Task 3 for Mathematical Practice 6.

*Visit **go.solution-tree.com/MathematicsatWork** for a free reproducible version of this figure.*

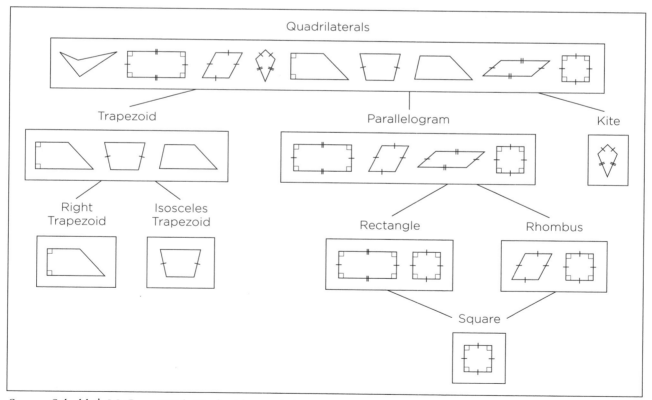

Source: Schuhl & McCaw, 2014. Used with permission.

Figure 6.21: Quadrilateral hierarchy.

Summary and Action

Standard for Mathematical Practice 6 requires students to attend to precision, whether by communicating effectively using appropriate vocabulary, using the correct units of measure, or calculating with the appropriate level of accuracy. Attending to precision also involves reflection and ensuring that solutions make sense in the context of the problem presented.

Identify a content standard you are having students learn currently or in the near future. Choose at least two of the Mathematical Practice 6 strategies from the following list to develop the habits of mind in students in order to attend to precision.

- Precise teacher language
  - Same Words—Different Meanings
  - Homophones
  - Measurement activities
- Precise student language
  - Back-to-Back
  - Reciprocal teaching
- Symbol literacy
- Mathematics vocabulary
  - Using vocabulary organizers
  - Creating active word walls
  - Playing games
- Estimation as a tool
- Error analysis
- Calculator feedback
  - Target Practice
  - Make It Happen

Record these in the reproducible "Strategies for Mathematical Practice 6: Attend to Precision." (Visit **go.solution-tree.com/MathematicsatWork** to download this free reproducible.) How were all students engaged when using the strategy? What was the impact on student learning? How do you know?

Chapter 7

Standard for Mathematical Practice 7: Look for and Make Use of Structure

Getting it *happens when students engage their minds in mathematical activity and connect what they are now learning to what they already know or to emerging structures in their minds.*

—CATHY SEELEY

While working in an elementary school, we happened to observe two lessons related to structure that showcased the need for Mathematical Practice 7, "Look for and make use of structure." In a first-grade classroom, we observed a teacher working with students to understand the commutative property of addition. Through investigation and repeated focused problems, students began to notice a pattern and articulate that the sum will be the same regardless of which number they write first in the expression. For example, 1 + 3 will be equal to 3 + 1. As this unfolded and students built on their own understanding of addition to make sense of the commutative property, the teacher acknowledged the structure and then proceeded to call it the *flip-flop property* because the numbers can be flip-flopped.

While students accepted the name and began to use it, we wondered, "What will happen when their second- or third-grade teacher uses the name *commutative property* instead? Will the student understand the structure of the property that works for all rational numbers or will this feel like a new concept since it has a new name?" Mathematical Practice 6 talks about the importance of precision with language, in part because it also helps students with Mathematical Practice 7.

In a fourth-grade classroom, we heard the following exchange.

Teacher: Find the value of 23 × 10. Explain your answer.

Marianna: It is 230. I know because I add a zero.

Teacher: Thank you, Marianna. Can someone else explain what she means by adding a zero?

Raphael: I think she means that she puts a zero at the end.

We must have had quizzical looks on our faces because the teacher asked us to join the conversation. Kit asked the class, "How many operations do we have in mathematics?" She gave the signal for a choral response, and we heard various versions of addition, subtraction, multiplication, and division. "So," she said, "do we have an operation called 'put a'?" The looks on students' faces were quite entertaining so she continued by asking "What happens when we add zero to a quantity?"

Roberto offered, "Nothing changes. The number stays the same."

Kit responded, "So, do we *add* a zero when we multiply a value by 10? Turn and talk with your partner. What happens when we multiply a value by 10?"

Students began discussing a variety of ideas, most of which involved expressing various ways to attach a zero to the end of the number. After a few minutes, Sarah drew a place-value chart and asked students to locate 23. They did so quickly, and we placed the digits in the appropriate places on the chart in figure 7.1.

Millions		Thousands			Units		
10	1	100	10	1	100	10	1
						2	3

Source: Adapted from Didax, 2013. Used with permission.

Figure 7.1: Place-value chart showing 23.

"What is the product when we multiply 23 × 10?" Sarah asked.

"230," they responded. Students added that value to the chart in figure 7.2.

Millions		Thousands			Units		
10	1	100	10	1	100	10	1
						2	3
					2	3	0

Source: Adapted from Didax, 2013. Used with permission.

Figure 7.2: Place-value chart showing 23 and 230.

"What happened to the 2? What happened to the 3? How has each digits' value changed as a result of multiplying by 10?" We asked students to continue to discuss this idea with their partners and then in groups of four. Their teacher then asked students to consider their new knowledge and asked, "If digits in a number have a value ten times greater when multiplying that number by 10, what happens to the digits' values when we multiply a number by 100?"

Students in this class began to change their language from "putting a zero" to the numbers shifting to a larger place value. They made the connection that two hundreds is ten times larger than two tens. Students in this fourth-grade classroom developed more precise language and deepened their understanding of place value.

Structure can also be found within the order of operations needed to simplify expressions. Figure 7.3 provides a sample of one student's work that incorporates the order in which operations must be addressed and calculated in an expression: (1) simplify expressions with parentheses and grouping symbols, (2) apply exponents, (3) multiply and divide expressions in order from left to right, and (4) add and subtract expressions in order from left to right. Many teachers use the acronym PEMDAS to describe the order of operations (parentheses, exponents, multiplication and division, addition and subtraction).

Grade 5 Task:

Charlie says $2 + 2 \times 2 = 8$. Gordon, however, says $2 + 2 \times 2 = 6$. Gordon is correct. What mistake did Charlie make?

Student Work

Charlie's mistake was he did the math problem straight across without doing Order of Operations or PEMDAs. Charlie did the addition first instead of multiplication.

PEMDAS

M goes before A

Figure 7.3: High-level task for Mathematical Practice 7.

*Visit **go.solution-tree.com/MathematicsatWork** for a free reproducible version of this figure.*

The student in figure 7.3 was able to explain the order of operations using PEMDAS and then clarifying the order using the words *addition* and *multiplication*. Unfortunately, the use of PEMDAS is often confusing for students because students do not understand the "left to right" idea for calculations. Too many students perform all of the operations as they read the expression from left to right rather than reading the entire expression and then applying the operations in order. As students learn operations, they learn to apply this order as a structure for correctly simplifying an expression.

There are many structures in mathematics such as those related to place value, numeric operations, and geometry. When understood, these lead to connections with future concepts, such as operations with fractions and algebraic thinking. The structures in mathematics reflect the consistency of mathematical reasoning. The symbols used in expressions and equations provide structure. The properties used to organize numbers and simplify expressions create structure. The shapes students learn in kindergarten and classify in fourth and fifth grade have consistent structure.

Structure lives in learning progressions as students build their understanding of number and shapes throughout their years in school and beyond. Explorations and explanations using structure lead to deeper conceptual understanding that can be consistently applied. This habit of mind focuses on creating new meaning from existing knowledge, which often develops students' abilities to reason deductively as well as inductively.

Table 7.1 shows what students do during a lesson to demonstrate evidence of learning Mathematical Practice 7 and what actions you can take to develop this critical thinking in students.

Table 7.1: Student Evidence and Teacher Actions for Mathematical Practice 7

	Student Evidence of Learning the Practice	Teacher Actions to Engage Students
Look for and make use of structure.	Students: • Identify structures in expressions • Identify structures in shapes • Connect previous concepts learned to new learning • Make a conjecture based on structure that eventually becomes a rule or algorithm • Compare and contrast application of new concepts to those previously learned • Explain why a property or algorithm works and when it should be used • Explain when a pattern can be generalized to apply to many cases and when it is specific to a particular problem • See and use structure in problems without being directed to do so	Teachers: • Provide opportunities for students to explore new concepts using previously learned structures • Ask questions that prompt students to make connections in their learning from one concept to the next and one year to the next • Create an anchor chart that lists the structures students have learned and should use • Have students share their thinking with one another • Provide tasks with patterns that lead to rules or algorithms • Connect big ideas • Make connections explicit with students • Avoid shortcuts and tricks and focus on understanding

*Visit **go.solution-tree.com/MathematicsatWork** for a free reproducible version of this table.*

Understand *Why*

When people talk of the beauty of mathematics, they are often referring to consistency within the structures of numbers, operations, and shapes. The addition strategies learned with single-digit whole numbers continue to work as an operation when the whole numbers students work with increase in size, become fractions and decimals, and later span all real numbers. For Mathematical Practice 7, looking for and making use of structure means students are making sense of mathematics.

What We Know About Mathematics Teaching and Learning (McREL, 2010) shows that "one way students learn is by connecting new ideas to prior knowledge. Teachers must help students come to view mathematics not as an isolated set of rules to memorize, but as the connection of ideas, mathematical domains, and concepts" (p. 27). Students learn as they are able to make connections based on an understanding of the mathematical structures they have previously learned. These learning progressions become the building blocks for future pattern recognition, exploration, and application.

Mathematics becomes much more difficult for students if they are taught skills in isolation. Kilpatrick and colleagues (2001) explain, "This practice leads to a compartmentalization of procedures that can become quite extreme, so that students believe that even slightly different problems require different procedures" (p. 123). This compartmentalization is reinforced when students are introduced to new topics using rules and algorithms rather than developing understanding that leads to generalizing and using rules and algorithms. What are the structures students can rely on when they encounter a problem so they do not randomly guess at a rule or algorithm to apply?

When introducing students to adding fractions, for example, have students make conjectures using models about how to add fractions based on their understanding of the structures learned for fractions as well as addition. This will eventually lead to algorithms for adding fractions. Students who learn the algorithm first may later find a common denominator when they are multiplying fractions because they are simply performing a procedure without understanding the task's structure.

Seeley (2014) further supports this assertion, stating:

> If a student memorizes a rule without *getting it*—understanding the mathematics behind the rule, when to use it, when it works, any limitations in using it, and so on—he or she may eventually forget or misunderstand all or part of the rule and may misapply it. Memorization can be a useful tool, but it's only part of *getting it*—the student needs to internalize what the rule is all about and recognize when it's helpful and when it's not. (p. 37)

The structures students learn related to place value allow them to understand regrouping when used in addition, subtraction, multiplication, and division. Students will explain their reasoning using concrete objects and models, pictures, expressions, and algorithms, flexibly using the method that is most efficient and accurate.

Kilpatrick and colleagues (2001) identify procedural fluency as the second interrelated strand that constitutes mathematical proficiency. They explain that this "refers to knowledge of procedures, knowledge of when and how to use them appropriately, and skill in performing them flexibly, accurately, and efficiently" (p. 121). Students develop procedural fluency as they learn the structure of numbers through an understanding of place value and rational numbers. This leads to an understanding of algorithms as structures that provide a "tool for completing routine tasks" (Kilpatrick et al., 2001, p. 121). Conceptual understanding will help students develop an understanding of mathematics, and this understanding of mathematics is critical for developing procedural fluency. This cycle shows the importance of building an understanding of mathematics through structure using connections.

In general, as Kanold (2012c) explains:

> Structure can help students learn what to expect in mathematics. If students learn how mathematics works and why it works the way it does, they then begin to notice, look for, and make use of structure to solve more difficult tasks and problems as they become engaged in what it means to *do* mathematics. (p. 43)

Why do students need this Mathematical Practice? Looking for and making use of structure means students are analyzing the consistencies in mathematics, deepening their own understanding of mathematics through connections, developing procedural fluency, and making generalizations to further explore mathematics and its applications. Students learn to think deductively and build new knowledge from previous understandings.

Strategies for *How*

What are the K–5 mathematical structures that students learn and use to frame their understanding from year to year? How do lessons identify these structures so students know which ones to look for? How do students recognize when they are using a previously learned structure?

When identifying those structures within a grade level or those that extend to many grade levels, consider the following questions.

- Is this concept a connection from previous learning (this year or a previous year)?

- Which previously learned structures can we use to investigate a new grade-level concept and extend it?

- How will students learn this concept later? What structures do I need to emphasize now for students to learn the concept more deeply later?

Table 7.2 offers a few mathematical concepts that lend themselves to deeper student learning through connections with structure.

Table 7.2: Examples of Mathematical Concepts Taught Through Structure

Domain or Strand	Structures
Counting and Cardinality	• Adding 1 to a number, resulting in the next number in the counting sequence • When counting objects, the last number said representing the number of objects in the group • Recognizing on a number line that values less than an identified number are to the left of the number while values to the right of the number are greater than the number

Operations and Algebraic Thinking	• Translating word problems to expressions and equations
	• Using properties such as:
	  • The associative property—$(3 + 4) + 5 = 3 + (4 + 5)$ or $(4 × 6) × 2 = 4 × (6 × 2)$
	  • The commutative property—$2 + 5 = 5 + 2$ or $7 × 6 = 6 × 7$
	  • The distributive property—$3 × (2 + 1) = 3 × 2 + 3 × 1$
	  • The identity property of addition—$5 + 0 = 5$
	  • The identity property of multiplication—$4 × 1 = 4$
	• Flexibly using inverse operations (addition-subtraction and multiplication-division)
	• Understanding that the less you add or subtract from a number, the closer the sum or difference will be to the original number

Domain or Strand	Structures
Number and Operations in Base Ten	• Using place value to compare and decompose numbers • Using place value to make sense of the addition, subtraction, multiplication, and division algorithms • Understanding operations with whole numbers and the connections to operations with fractions and decimals
Number and Operations— Fractions	• Using unit fractions to compose and decompose fractions • Understanding a denominator and numerator in a fraction and their relationship to a whole • Understanding operations with fractions and when common denominators are needed • Comparing and estimating fractions using benchmark fractions on a number line
Measurement and Data	• Applying length, area, and volume from an understanding of a collection of units, square units, and cubic units without gaps or overlaps • Applying operations in the context of measurements, money, or collected data • Using a clock to tell time and solve problems related to time • Connecting quarter hours to unit fractions
Geometry	• Identifying and classifying two- and three-dimensional shapes using measurable attributes

In the standards, students learn several structures. Though still important, some structures may later be de-emphasized as students continue to use them. They can still be included in a rich task for further understanding. For example, addition and subtraction are emphasized in grades K–2, but by third and fourth grade, students use that learning to become fluent with the standard algorithm for addition and subtraction and apply the operation to larger numbers and, eventually, fractions and decimals.

There are structures related to content, but there are also structures related to reasoning and solving problems. It's important to consider whether students recognize when tasks pose a similar structure and for them to have an idea of how to best solve the task at hand. The strategies for this Mathematical Practice related to looking for and making use of structure will address the structures in content and reasoning. Building students' capacities to identify and understand structures is the goal. The challenge will be to have students understand the structures well enough that *they* see them and look for them on their own. Classroom activities for building this understanding may involve using Brain Splash strategies, algebraic thinking, connections, inferences from structure, and complex problems.

 ## *Brain Splash*

Once students have spent time learning grade-level concepts, Brain Splash strategies become useful tools for having students make connections between previous learning and current content. Each of the Brain Splash strategies provides an opportunity for students to show what they currently know about a concept based on the structures of the concept they have previously learned. They also serve as diagnostic tools. Analyzing students' work provides insights into their thinking and helps you plan next steps. Provide opportunities for students to make connections and "get it" through working with others and experiencing the consistency of mathematics.

KWL

KWL is a well-known literacy strategy that also works for mathematics. In first grade, for example, you might ask students what they already know (K) about addition and subtraction prior to any instruction with those operations. Students might draw pictures or write equations. You might also make it an oral activity and generate responses on chart paper. Next, students generate questions about what they want (W) to know or what they need clarification on based on prior learning. Students can write their responses or share them with an elbow partner and then with the whole group. Finally, after instruction, revisit the chart, and have students identify what they have learned (L). See figure 7.4 for a sample chart.

What Do I Know?	What Do I Want to Know?	What Did I Learn?

Figure 7.4: KWL chart.

*Visit **go.solution-tree.com/MathematicsatWork** for a free reproducible version of this figure.*

In the intermediate grades, if students have written what they know and want to know related to fractions, for example, you might also incorporate a Give One, Get One strategy. Have students share with a partner what they know, and allow the partner to add to his or her chart ideas that they have already learned. Next, students trade places. You can then have each student repeat the exercise with a new partner or share some of the responses as a whole class.

When students complete what they know and what they want to learn, it allows teachers to see any preconceived ideas they have related to the concept. Also, it allows teachers to see any possible misconceptions students might have so they can be addressed as soon as possible. Most importantly, it helps a student build new learning from previous learning.

Figure 7.5 shows a fourth-grade student's KWL chart on fractions.

What Do I Know?	What Do I Want to Know?	What Did I Learn?
• That there is a denominator and a numerator • That the numerator is the shaded amount • That the denominator is the total amount	• Why do fractions go up to the biggest number? • Why are fractions so important to the world? • Why don't younger kids learn fractions?	• Fractions can be simplified • Fractions can be equivalent • Fractions can be added • Fractions can be subtracted

Figure 7.5: Sample fourth-grade fraction KWL chart.

Teachers can also modify the KWL to a KEL chart. In this case, the *E* stands for *explore*, and after students write what they know, they explore a problem using their knowledge before instruction begins. After instruction (that day or a later day), students complete the column explaining what they learned.

Journal Prompt

Students can show their previous knowledge and relate it to current material with a quality journal prompt. You might even allow students to work on their prompt in pairs or groups if the intent is to activate prior knowledge and explore its connection to current content. When journals formatively assess a student's understanding of structure, the response must include shown work and an explanation. Figure 7.6 provides sample prompts and explanations for using them.

Grade	Example Prompt	Why This Prompt?
K	Show two ways to find the answer to 3 + 5.	Use this prompt after students have worked on sums within 5. Pose this question to see if they can use the structure of learned addition strategies (for example, number line, ten frame, objects) to find 3 + 5. Can they extend the structure of addition and addition strategies beyond a sum of 5?
1	How many different ways can you write 27 using tens and ones? Show each way.	Students who write 2 tens and 7 ones, 1 ten and 17 ones, and 27 ones show an understanding of place value beyond recognizing only a single digit as a way to represent place value (2 tens and 7 ones).
2	When you start at 0 and skip-count by twos, you identify even numbers. Mia says that means 35 is an even number. Is she correct? Explain how you know.	Use this prompt after students know how to skip-count by twos. Can they apply the structure definition of even numbers to their knowledge of skip-counting?

continued →

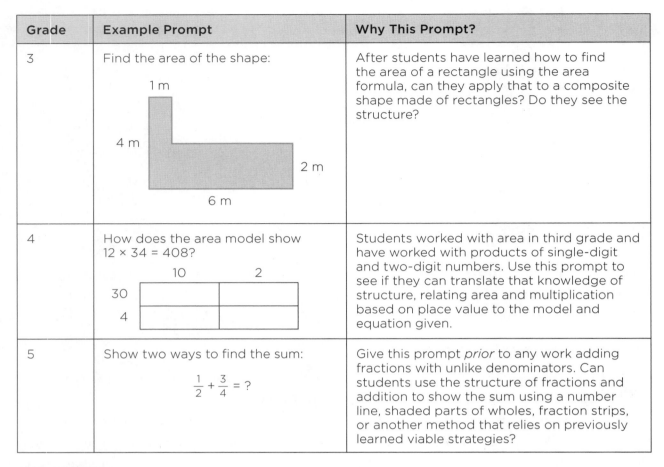

Grade	Example Prompt	Why This Prompt?
3	Find the area of the shape: 1 m 4 m 2 m 6 m	After students have learned how to find the area of a rectangle using the area formula, can they apply that to a composite shape made of rectangles? Do they see the structure?
4	How does the area model show $12 \times 34 = 408$? 10 2 30 4	Students worked with area in third grade and have worked with products of single-digit and two-digit numbers. Use this prompt to see if they can translate that knowledge of structure, relating area and multiplication based on place value to the model and equation given.
5	Show two ways to find the sum: $\frac{1}{2} + \frac{3}{4} = ?$	Give this prompt *prior* to any work adding fractions with unlike denominators. Can students use the structure of fractions and addition to show the sum using a number line, shaded parts of wholes, fraction strips, or another method that relies on previously learned viable strategies?

Figure 7.6: Sample journal prompts by grade.

Exploration

When you are teaching a concept reliant on prerequisite skills students learned earlier in the year or in a previous grade level, consider how to pose the concept as an exploration. Given a quality task to start the lesson, students can show their current level of understanding as they apply what they know.

For example, in third or fourth grade, provide a task requiring addition or subtraction within 1,000 (third) or 1,000,000 (fourth) to see how they apply their understanding of addition to larger numbers than they used in a previous grade. When discussing the solution to the task, help students see how they can use their knowledge about the structure of place value when adding and subtracting to add and subtract any number. This extends to decimals as well. Allow students the opportunity to realize they have learned the concepts they need to solve the task at hand.

Additionally, when students are learning about converting from smaller units to larger units in fifth grade, provide an exploration that has them first review what they have learned related to conversions from larger units to smaller units. Then, see if they can use inverse reasoning, which may include their understanding of the structure of multiplication and division as inverse operations, to find the conversions from smaller to larger units. Figure 7.7 offers an example.

Metric Conversions

Write the missing value in each conversion. Show the equation you used or explain how you found each missing value.

1. 1 centimeter = _____ millimeters
2. 3 centimeters = _____ millimeters
3. 1 meter = _____ centimeters
4. 4 meters = _____ centimeters
5. 1 kilometer = _____ meters
6. 6.2 kilometers = _____ meters
7. 2 meters = _____ millimeters
8. 9.5 kilometers = _____ centimeters

Based on the information in problems 1 through 8, and your knowledge of conversions, write the missing value in each conversion. Show the equation you used or explain how you found each missing value.

9. _____ centimeters = 20 millimeters
10. _____ centimeters = 65 millimeters
11. _____ kilometers = 4,000 meters
12. _____ meters = 740 centimeters

13. How would you tell a friend to convert from a larger unit to a smaller unit, as you did in numbers 1 through 8?

14. How would you tell a friend to convert from a smaller unit to a larger unit, as you did in numbers 9 through 12?

Figure 7.7: Exploration for metric conversions.

Visit **go.solution-tree.com/MathematicsatWork** *for a free reproducible version of this figure.*

Alternately, you might give just one problem, such as the following.

> Lori measured the length of the hallway at school to find the length of ribbon needed to decorate the wall. She measured the hallway and found it was 4,250 centimeters long. The ribbon is measured in meters, and there are 8 meters in each package. When she tapes the ribbon along the hallway, she can place the end of one piece to the start of the next piece and continue down the hallway. How many packages of ribbon does she need to buy to make sure she has enough ribbon for the length of the hallway? Use words, numbers, and/or pictures to justify your answer.

Some students will make sense of the task through a picture. Some will multiply and find the product of 8 meters per package and 100 to find the centimeters of ribbon in each package and then use that information to find the total number of packages needed. Other students will divide 4,250 centimeters by 100 to find the number of meters or ribbon needed and then determine the number of packages. Have students share their different strategies, and then ask students how they connect so they see the structure inherent in conversions and can apply that structure to future tasks.

Algebraic Thinking

Much algebraic thinking is based on the structure of numbers and operations. This can translate into mental mathematics as well as conceptual understanding that students can apply to new topics.

Mental Mathematics

When students mentally add, subtract, multiply, and divide, they rely on the structures of number, place value, and operations to efficiently and flexibly perform the computations. The goal is for students to use mental computation. See figure 7.8 for examples.

Grade	Problem	Mental Computation	Explanation
K	$1 + 3$	$3 + 1 = 4$	Use the commutative property to reverse the addends.
1	$14 + 9$	$14 + 10 = 24$ $24 - 1 = 23$	Compose 9 as $10 - 1$.
2	$45 - 27$	$45 - 20 = 25$ $25 - 7 = 18$	Decompose 27 as 20 and 7.
		$45 - 5 = 40$ $40 - 20 = 20$ $20 - 2 = 18$	Decompose 7 as 5 and 2.
3	5×14	$5 \times (10 + 4) = 5 \times 10 + 5 \times 4$ $= 50 + 20$ $= 70$	Use the distributive property based on place value.
4	$\frac{1}{8} + \frac{5}{8}$	1 one-eighth plus 5 one-eighths = 6 one-eighths $\frac{6}{8}$ or $\frac{3}{4}$	Think of the number of unit fractions being added to find the total number of unit fractions in the sum.
5	$1\frac{1}{3} + \frac{3}{4} - \frac{1}{3} + 4\frac{1}{4}$	$1\frac{1}{3} + \frac{3}{4} - \frac{1}{3} + 4\frac{1}{4} = (1\frac{1}{3} - \frac{1}{3}) + (\frac{3}{4} + 4\frac{1}{4})$ $= 1 + 5$ $= 6$	Use the associative property to reorder the values in order to group like denominators.

Figure 7.8: Examples of mental mathematics strategies.

Consider using mental mathematics problems outside of the mathematics block of time as well as during a mathematics lesson. Pose a question orally or in writing, and have students sit on their hands (no writing) and think for one minute about how many ways to solve the problem they think they can share. Next, randomly call on students to share how they thought about the solution and record their thinking for all students to see the various strategies. Emphasize the structure of number, place value, or operations that students use when thinking about computations mentally. You might also connect their thinking to previously learned strategies to include using concrete objects, number lines, or area models through questioning.

Comparisons

In the primary grades, students learn that 31 is greater than 24 because three tens is greater than two tens, regardless of the ones that follow, due to the structure of place value. They also learn from the structure of composing numbers that 3 + 2 is greater than 3 + 1. These ideas translate into algebraic comparisons by having students determine the greater of two values as shown in figure 7.9. When comparing the expressions, students use structure to determine the answer without filling in values for the missing numbers.

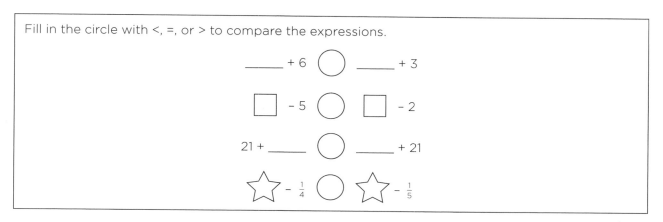

Figure 7.9: Algebraic comparison examples.

You can create a game with this algebraic thinking by having students create cards or giving students previously created cards to use to play a game with a partner, much like the card game War. Each student flips over a card, and the student whose expression is greater wins both cards. The winner is the person with the most cards at the end of a round or, if the rounds continue, eventually wins all of the cards. Be sure when cards are made that every card has the same symbol (line, rectangle, star, variable, and so on) in the expression so students can accurately compare their card, regardless of when it is flipped over.

Connections

When thinking through connections in mathematics, help students make connections from previous learning to current learning by referring to previous solution strategies or rules they have learned and applied. This will emphasize the consistent structure that can be used in mathematics. The following strategies help students make connections to solutions or rules and between concepts.

Connecting Solution Pathways

When posing a task, ask students, "How does this task seem similar to another task you have solved before?" Allow students to identify content from other tasks throughout the year or from a previous grade that they feel resembles this task, and have them explain why. Help students make connections between the tasks by asking questions like the following.

- "How did you solve the previous task?"

- "Will that strategy work to solve this task?"

- "What ideas in this task are similar to the previous task?"

- "What in this task is different or new? How will you deal with that in your solution?"

- "What skills and strategies have you learned that might help you solve this task?"

- "What do you wonder about in this task?"

Students should look for structure that they can apply to solve future tasks. When students are developing this habit of mind, the connections between solution pathways are another structure to emphasize with students. For example, students might remember they used a picture, a number line, or an equation to solve a similar task and will therefore try that strategy again if it was successful the first time.

Connecting Situations to Rules

Mathematical algorithms, definitions, and theorems are rules that K–5 students learn that show the structures students should be looking for and using when solving tasks. Remember, students must first make sense of concepts before generating the rule or algorithm. It is important that students understand when they can and cannot apply the rules.

For example, when students are solving a subtraction problem like $27 - 18$ using a standard algorithm, many students will say "Well, you can't subtract a larger number from a smaller number, so I can't subtract 8 from 7." But in fact, you *can* subtract a larger number from a smaller number ($7 - 8 = -1$), which students will learn in later grades. It's important that rules students learn early on (such as the standard algorithm for single-digit subtraction) prepare them for later rules (such as the standard algorithm for double-digit subtraction or for negative numbers). Beware the absolutes like "*never* subtract a larger number from a smaller number." In K–5, students work with numbers greater than or equal to 0. Consider asking students where they would land on a number line if they jump back eight places from 7 based on their structural understanding of moving left when subtracting. Discuss the mirror image to the positive numbers called negative numbers. You might even show when subtracting $27 - 18$ that 2 tens minus 1 ten is 1 ten, and 7 ones minus 8 ones is –1 one, and so the final answer is the sum of 10 and –1 which is 9. There is not a need to spend much time on this (though some students will be curious and can explore negative numbers further). In seventh grade, students will do this easily. Remind K–5 students that when they subtract, they need all of their answers to be greater than or equal to zero.

The order of operations provides another opportunity to be clear about structure. Display student work that shows a common misconception, and have students in the class identify the mistake and fix the problem from there. Alternately, you can identify the most common incorrect answer and ask groups of students what mistake a student made to get that incorrect answer. This requires complex reasoning and promotes quality discussions while deepening student understanding.

Another area that lends itself to being clear about rules and definitions is geometry. A rectangle, for example, is defined as a quadrilateral with four right angles. This means a square is a special rectangle since a square also has four right angles. A square, however, has more to its definition that includes having four congruent sides. Have students focus on the definitions (structure) of shapes to determine how they are related to one another. Figure 7.10 shows one example. Other geometry examples are given in the lesson for Mathematical Practice 6 on page 169.

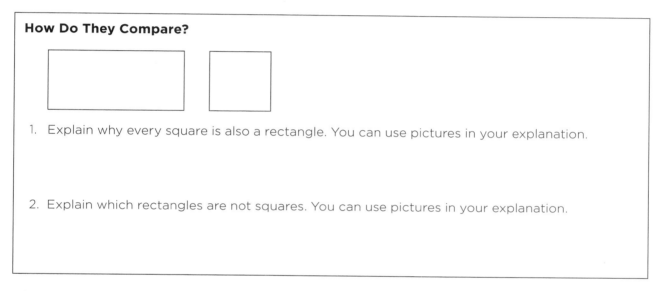

Figure 7.10: Geometry structure example.

*Visit **go.solution-tree.com/MathematicsatWork** for a free reproducible version of this figure.*

When clarifying the conditions for algorithms, definitions, and theorems as rules, consider asking students questions like the following.

- "Is that always true?"

- "Will that always work?"

- "When will the rule work? When will it not work?"

- "When is that rule needed?" (For example, when should students decompose when subtracting? When should students find a common denominator when working with fractions?)

- "When do you think you will use this algorithm? Is it always the most efficient strategy?"

What Doesn't Belong? from chapter 3 (page 70) can also help students understand structure. Give students four expressions, equations, drawings, shapes, tools, or other mathematical representations, and have them identify which of the four does not belong with the others. They must then explain their reasoning. This activity most effectively provides an opportunity for students to connect the dots related to structure when there is more than one right answer. Students can also create their own What Doesn't Belong? board for other students to analyze and discuss. See figure 7.11 for an example.

What Doesn't Belong?

50 + 50	5 + 5
10 × 10	8 + 8

Figure 7.11: What Doesn't Belong? example.

Students might give the following responses based on their understanding of each expression's structure, all of which are correct and viable (Mathematical Practice 3).

- 8 + 8 because it is the only expression that does not have an answer that is a multiple of 10.
- 10 × 10 because it is the only expression that involves multiplication.
- 5 + 5 because it is the only expression with two odd numbers.
- 50 + 50 because it is the only expression with numbers greater than 10 in the expression.

Inferences From Structure

Students can make conjectures and inferences from structure—inferences about task solutions, geometric patterns, and algorithms. Each inference is based on deductive reasoning, leading to conclusions and possibly new learning.

Standard Algorithms

Students must first understand a concept like addition, subtraction, multiplication, or division before they can establish fluency or proficiency with an algorithm. As students learn different strategies, always emphasize place value, which eventually leads to their ability to understand and use standard algorithms. Also, remind students to choose the most easy, efficient, and effective method (see the Three Es strategy in chapter 4, page 106). Write the different representations on chart paper to collect the various methods, and link them to the algorithm.

For example, consider addition. See figure 7.12. Once students have added using concrete objects and ten frames, see the structure of adding tens and ones, and are adding multidigit numbers, they will use an open number line and base-ten blocks.

Add Tens and Then Ones	Add Ones and Then Tens (Not the Standard Algorithm)	Standard Algorithm for Addition
25 + 68 = ? 20 + 60 = 80 5 + 8 = 13 80 + 13 = 80 + 10 + 3 = 93 or	25 + 68 = ? **Partial Sums:** $$\begin{array}{r} 25 \\ + 68 \\ \hline 13 \\ + 80 \\ \hline 93 \end{array}$$ or	25 + 68 = ? $$\begin{array}{r} 1 \\ 25 \\ + 68 \\ \hline 93 \end{array}$$

(Table continued)

Tens / **Ones** table: 	Tens	Ones
---	---
2	5
+ 6	8
8	13 | = 8 tens + 13 ones = 8 tens + 1 ten + 3 ones = 9 tens + 3 ones = 93	**Add the ones and write the resulting ten below the numbers being summed and above the line.** $$\begin{array}{r} 25 \\ + 68 \\ \hline 1 \\ \hline 93 \end{array}$$	

Figure 7.12: Addition learning progression from understanding to algorithm.

The nonalgorithmic models showing flexibility with adding tens before ones lead to stronger mental mathematics and a deeper understanding of structure. Once students learn the standard algorithm, they can use the structure in tasks. The standard algorithm is the last structure when learning the remaining operations.

Patterns With Structure

You can have students compose and decompose geometric shapes based on their understanding of shapes' structures. They can then also make inferences about other shapes based on their learning. For example, for kindergarten, give students pattern blocks and ask them to make the following.

- Use two triangles to make a rhombus.
- Use three triangles to make a trapezoid.
- Use two trapezoids to make a hexagon.
- How many triangles are in a hexagon?
- How many rhombi are in a hexagon?

For grade 3, provide tangrams as in figure 7.13 (page 194), and ask students to compose a shape using the tangram pieces.

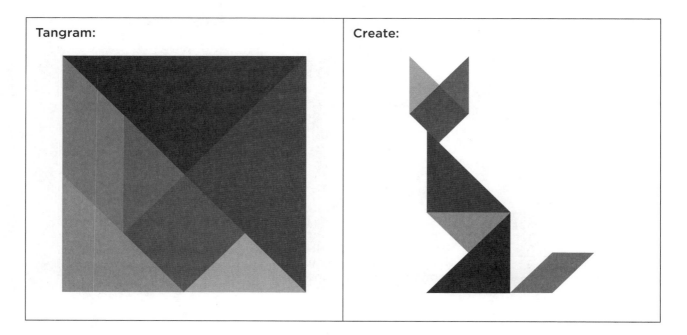

Figure 7.13: Tangrams.

*Visit **go.solution-tree.com/MathematicsatWork** for a free reproducible version of this figure.*

You can also ask students to find the area of each tangram piece if the area of the outer square is one square unit. Students will reason through the structure of separating the square into equal-sized pieces and finding the fractional area of each based on the number of pieces of that size that could be inside the tangram. Look for students to check their work by summing the fractional areas to make sure they have a total of one square unit. They can determine the relationships between the shapes by stacking, for example, two small triangles on the square, and so on.

Complex Problems

When students are given a complex task, they must make a solution plan that breaks the problem into smaller tasks to complete. Your job is to provide complex tasks and have students discuss their strategies, showing the smaller tasks they solved along the way toward the final solution. Examples for grades 3 and 4 follow.

> ### Grade 3
> In a 24-hour period, for how many minutes does at least
> one 5 appear as a digit on the clock?

Students solving this task will have to reason through several pieces of information. In some order, their response will include the following three steps.

1. Count how many minutes on the clock show a 5 in the ones place: 0:05, 0:15, 0:25, 0:35, 0:45, 0:55—there are 6 minutes with a 5 in the ones place.

2. Count how many minutes on the clock show a 5 in the tens place: 0:50, 0:51, 0:52, 0:53, 0:54, 0:55, 0:56, 0:57, 0:58, 0:59—there are 10 minutes with the 5 in the tens place, but 0:55 was counted in the ones place already.

3. Count the minutes in the two hours showing a 5 because the hour is 5 (a.m. and p.m.)—there are 120 minutes.

There are 22 hours with 6 + 10 − 1 = 15 minutes showing a 5, for a total of 22 × 15 = 330 minutes. Also, there are 2 hours with 120 minutes showing at least one 5. This means in a 24-hour period, a 5 is shown at least once for a total of 330 + 120 = 450 minutes or 7.5 hours.

Students will break the task into chunks that make sense to them based on the structure of how time works—with hours and minutes—in a twenty-four-hour period of time.

Grade 4

A pizza store makes two sizes of pizzas shaped as rectangles.
The large pizza is twice as long and twice as wide as the small pizza.
Fadia ate $\frac{3}{4}$ of a small pizza, and Lori ate $\frac{3}{16}$ of a large pizza.
Fadia says she ate more pizza because $\frac{3}{4}$ is greater than $\frac{3}{16}$. Is she correct?
Explain your reasoning.

This task requires students to compare the sizes of two different wholes for each pizza. Most students will find a graphic representation of the pizza Fadia ate and a separate graphic representation for the amount of pizza Lori ate in order to compare the fractions and see that Fadia and Lori ate the same amount of pizza. Figure 7.14 shows an example for how to compare the total square units even with different-sized wholes.

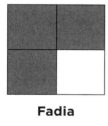

Fadia

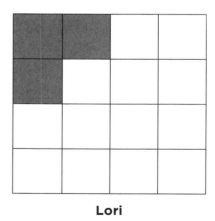

Lori

Figure 7.14: Graphic representation of grade 4 pizza task.

As students work through high-level tasks requiring separation into smaller problems to solve, consider asking questions like the following for students who are stuck.

- "What is the task asking you to find?"

- "What do you know? What do you need to know?"

- "Can you draw a picture to help you? What could the side lengths measure on each pizza?"

- "What is one thing you think you need to do to solve this task?"

- "How does this task resemble something we have been working on in class?"

When giving students complex tasks to solve, it is important to not begin by telling students what to do first, what to do second, and so on. This turns a high-level task into several low-level tasks. Instead, strategically group students and have students productively struggle to find the smaller chunks they need to solve in order to have a reasonable solution. You may need to do a minilesson with a group of students but not with the entire class.

Lesson Example for Mathematical Practice 7: Connecting Rectangles and the Distributive Property

The lesson plan in figure 7.15 focuses on looking for and making use of structure. It shows the connection between the structure of rectangles and the distributive property. Prior to this lesson, students have learned that area is the number of unit squares that completely cover a two-dimensional shape with no gaps or overlaps. They have made connections between multiplication in an array and in a rectangle tiled with unit squares. This led to an exploration that generated the formula for the area of a rectangle as *area = length × width*. In a previous unit, students have also learned that multiplying a single-digit number by a multiple of 10 results in the original digit shifting one place value to the left.

Students will be able to take the structures learned in this lesson related to combining areas and the distributive property to further explore finding the area of composite shapes composed of rectangles and using an area model to find the product of a single-digit number and a two-digit number using place value. Although Mathematical Practice 7 is the focus of the lesson, students will also have to make sense of problems (Mathematical Practice 1), construct viable arguments and critique the reasoning of others (Mathematical Practice 3), and be precise with notation when writing expressions and equations and drawing rectangles (Mathematical Practice 6). A commentary follows the lesson providing more information related to the rationale and importance of each lesson component. Figures 7.16 and 7.17 (pages 199–201) support the lesson's tasks.

Unit: Area

Date: March 10

Lesson: Connect area models to the distributive property. (3.MD.7c)

Learning objective: As a result of class today, students will be able to explain why the distributive property works using area models.

Essential Standard for Mathematical Practice: As a result of class today, students will be able to demonstrate greater proficiency in which Standard for Mathematical Practice?

Mathematical Practice 7: "Look for and make use of structure."

- Students will use the structure of rectangular area to generate the distributive property.
- Students will use the distributive property to find the product of a single-digit number and a two-digit number.

Formative assessment process: How will students be expected to demonstrate mastery of the learning objective during in-class checks for understanding teacher feedback and student action on that feedback?

- Students will make thinking visible using drawings and equations. This will allow the teacher to give meaningful descriptive feedback to students while they are working in groups.
- Students will share their thinking with partners and the whole class and then continue their work based on that feedback.

Probing Questions for Differentiation on Mathematical Tasks

Assessing Questions	Advancing Questions
(Create questions to scaffold instruction for students who are stuck during the lesson or the lesson tasks.)	(Create questions to further learning for students who are ready to advance beyond the learning standard.)
• What do you know? • How can you use grid paper to show your thinking? • How can you use cubes to show your thinking? • What is the formula for the area of a rectangle? Why does it work?	• What is 8×32? • What is 8×132? • Show three different ways to draw an area model showing 4×12. Which one makes the most sense to you? Why?

Tasks (Tasks can vary from lesson to lesson.)	What Will the Teacher Be Doing? (How will the teacher present and then monitor student response to the task?)	What Will Students Be Doing? (How will students be actively engaged in each part of the lesson?)
Beginning-of-Class Routines How does the warm-up activity connect to students' prior knowledge, or how is it based on analysis of homework?	Ask students to find each product: 1. 2×10 2. 5×10 3. 9×10 4. 4×10 Share the answers. Have students talk with a partner about how they solved 4×10. Next, randomly call on a person to explain how he or she found the product of 4×10. Invite others to share their reasoning too. Ask students to draw a rectangle with side lengths of 4 and 10 and find its area. How is the area related to 4×10?	On a whiteboard or piece of paper, students write the product for each expression. Students talk with a partner to articulate how they found each product. A few students share their thinking with the class while others listen and agree or disagree. Students draw a rectangle and write its area. They discuss how the area formula is the product of length and width, which matches 4×10.

continued →

Task 1 How will students be engaged in understanding the learning target? (See figure 7.16.)	Have students work in pairs to complete the task. Walk around the room and monitor student work and progress, providing feedback, if needed. Ask students to share their inferences, and lead students to the distributive property. Be sure students notice the following. • All three rectangles have one side length in common. • The remaining side lengths of A and B sum to the other side length of C. • If you put rectangles A and B together, they would be congruent to rectangle C. On the board or piece of chart paper, as students discuss, write: 1. $(2 \times 6) + (2 \times 3) = 2 \times 9$ 2. $(4 \times 3) + (4 \times 7) = 4 \times 10$ 3. $(8 \times 1) + (8 \times 2) = 8 \times 3$ Let them know this is the *distributive property* and can be shown using area of rectangles.	Students work in pairs to complete the task. They make observations and inferences about the relationship of area of the rectangles when separated and put together. Share answers and work collectively as a class to "see" the distributive property.
Task 2 How will the task develop student sense making and reasoning? (See figure 7.17.)	Have students work in pairs to complete the task. Select students with different representations for question 2 to share their work with the class so students see there is more than one way to separate the larger rectangle with an area of 40 into two smaller rectangles.	Students work in pairs to determine the answers to the task. They share their answers with others.
Task 3 How will the task require conjectures and communication?	Ask students to find 5×12 using the distributive property. Have students share their whiteboards with the class for different representations. Emphasize a representation that shows $(5 \times 10) + (5 \times 2)$, which is built using place value. This will help students as they begin to explore multiplying a single digit by a two-digit number and multiplying a single digit by a multiple of 10. Other representations are also possible. If there is time, show the following rectangle, and ask students to write an expression for its area. $6 \begin{array}{\|c\|c\|} \hline & \\ \hline \end{array}$ $\quad 10 \qquad 7$	Students work with a partner to make a rectangle that is 5 by 12 and then decompose the 12 into two values. They find the area of each smaller rectangle and sum them to find 5×12. Students write 6×17 or $(6 \times 10) + (6 \times 7)$.

Closure		
How will student questions and reflections be elicited in the summary of the lesson? How will students' understanding of the learning target be determined?	Teacher poses one or more of the following questions and listens to student responses. • Do you prefer to find the area of the outside rectangle or find the sum of the two inside rectangles when finding the area of a large rectangle? Why? • How were the tasks today similar to tasks you have done before? • What did you learn today? • What are you wondering about?	Students answer the questions with a partner and then share their responses (when randomly called) with the class.

Source: Template adapted from Kanold, 2012c. Used with permission.

Figure 7.15: Grade 3 lesson-planning tool for Mathematical Practice 7.

*Visit **go.solution-tree.com/MathematicsatWork** for a free reproducible version of this figure.*

Task 1

Name: _____   Date: _____

Distributive Property Investigation

Complete the chart. What do you notice?

Area of Each Rectangle	Area of A + Area of B	Area of C
2 [A] 2 [B] 3 6 2 [C] 9 Area of A = _____ square units Area of B = _____ square units Area of C = _____ square units		
4 [A] 4 [B] 3 7 4 [C] 10 Area of A = _____ square units Area of B = _____ square units Area of C = _____ square units		

continued →

Area of Each Rectangle	Area of A + Area of B	Area of C

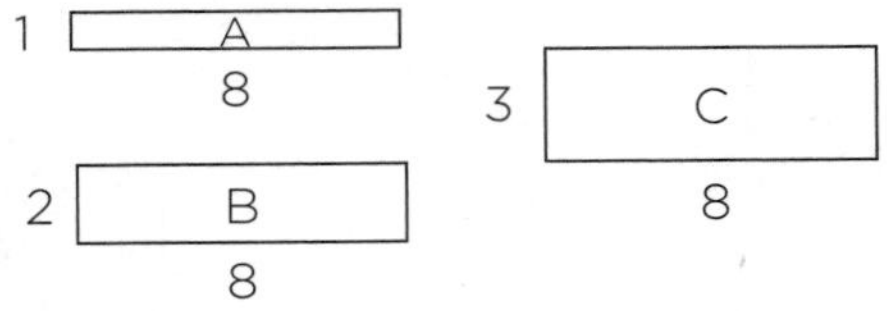

1 ☐——A——☐
      8
              3 ☐   C   ☐
                    8
2 ☐   B   ☐
      8

Area of A = _____ square units
Area of B = _____ square units
Area of C = _____ square units

What do you notice about the three rectangles in each problem?

What do you notice about the area of A + area of B and the area of C?

Why do you think this pattern works for these three problems? What must be true about three rectangles for this pattern to work?

Figure 7.16: Task 1 for Mathematical Practice 7 grade 3 lesson.

*Visit **go.solution-tree.com/MathematicsatWork** for a free reproducible version of this figure.*

Use the beginning-of-class routine as an opportunity to review students' previous learning related to multiplying a single digit by 10 and finding the area of a rectangle using a formula. If students struggle, build a 4 × 10 rectangle using four base-ten blocks to show an array, and then students can skip count by 10 to find the product and area of forty square units. Emphasize the length units versus square units for area.

Task 1 is designed to have students look for and make use of the structure of a rectangle's area, which they have previously learned, and its connection to multiplication and addition. From this, teachers can consider whether students generate the structure of the distributive property.

Task 2 solidifies the learning from the class's conjecture. Students become clearer about how they can use the distributive property to find areas of rectangles. Task 3 then introduces how these distributive property models will help them find products of single-digit and two-digit numbers using place value.

Task 2

Name: _____   Date: _____

Distributive Property With Rectangles

1. Show two different ways to find the area of the outside rectangle.

3 in.

4 in.   5 in.

Area of 2 Small Rectangles Added Together	Area of 1 Large Rectangle Using One Equation

2. Draw a picture to show two rectangles side by side that together have an area of 40 square inches. Explain how you know your answer is correct using equations and words.

Figure 7.17: Task 2 for Mathematical Practice 7 grade 3 lesson.

*Visit go.**solution-tree.com/MathematicsatWork** for a free reproducible version of this figure.*

Students close with an explanation of what they learned related to the distributive property and connect that knowledge to some previous tasks and learning related to multiplication and area of rectangles. Emphasize the connections that occurred through structure.

Summary and Action

Standard for Mathematical Practice 7 requires students to look for and make use of structure. This means they must have an understanding of the structures inherent to place value, algebraic thinking, numeric operations, and geometry. Additionally, they must understand the structures they use when solving tasks. Students who develop this habit of mind make connections to current concepts using information from concepts they have previously learned. They also apply their knowledge to new situations because they understand the consistency of the structures.

Identify a content standard you are having students learn currently or in the near future. Choose at least two of the Mathematical Practice 7 strategies from the following list to develop the habits of mind in students in order to look for and make use of structure.

- Brain Splash
  - KWL
  - Journal prompt
  - Exploration
- Algebraic thinking
  - Mental mathematics
  - Comparisons
- Connections
  - Connecting solution pathways
  - Connecting situations to rules
- Inferences from structure
  - Standard algorithms
  - Patterns with structure
- Complex problems

Record these in the reproducible "Strategies for Mathematical Practice 7: Look for and Make Use of Structure." (Visit **go.solution-tree.com/MathematicsatWork** to download this free reproducible.) How were all students engaged when using the strategy? What was the impact on student learning? How do you know?

Chapter 8

Standard for Mathematical Practice 8:
Look for and Express Regularity in Repeated Reasoning

To grow mathematically, children must be exposed to a rich variety of patterns appropriate to their own lives through which they can see variety, regularity, and interconnections.

—LYNN ARTHUR STEEN

Recently, Kit had an opportunity to work with Courtney, a second-grade student. Courtney was asked to find the sum of 43 + 29. Figure 8.1 illustrates how her teacher presented the problem emphasizing the use of the traditional algorithm.

$$
\begin{array}{r|r}
4 & 3 \\
+\ 2 & 9 \\
\hline
\end{array}
$$

Figure 8.1: Place value example.

Courtney was stuck in the "counting on" strategy for each place value. She used a hundreds chart and started at 9 and then counted 3 more to find the sum of the numbers in the ones column. She had not considered any patterns that might make finding totals easier. Kit began by asking her to find the sum of 10 + 2. She looked at her hundreds chart and counted two more from ten. Kit asked her what she noticed about the sum 10 + 2 = 12 and recorded the following number sentences as Courtney found other sums.

- 10 + 2 = 12
- 10 + 4 = 14
- 10 + 5 = 15

Courtney said, "Wait, that number is in the chart. I don't have to count." Just to check, Kit asked her about 10 + 8, 10 + 7, and so on.

Courtney recognized the pattern and found the other sums starting with 10 easily. She started to demonstrate Standard for Mathematical Practice 8, "Look for and express regularity in repeated reasoning."

Kit then asked her about 9 + 3, and showed her two columns of Unifix cubes. Courtney asked, "Can I make a ten? That is easy." She then extended her thinking to other sums and said, "This works for any values close to 10 or 20 or 30."

Kit probed further.

Courtney said, "I can make numbers that are easier to use than the ones in the problem without changing the total. I can do this with any values."

Courtney has now extended her thinking to create a generalization about equivalent expressions. Her thinking is again demonstrating Mathematical Practice 8. As her mathematical thinking matures, Courtney will be able to prove her generalization and extend it as she engages in Mathematical Practice 7, "Look for and make use of structure."

The Standards for Mathematical Practice frequently intertwine. As the preceding scenario illustrates, Courtney engaged in both Mathematical Practices 7 and 8. As both of these practices involve working with patterns, it can be difficult to identify one from the other. Mathematical Practice 7 emphasizes structure, and Mathematical Practice 8 emphasizes patterns through repeated reasoning.

Table 8.1 explains some of the distinguishing features between these two practices.

Table 8.1: Comparison of Mathematical Practices 7 and 8

Mathematical Practice 7 When students look for and make use of structure, they . . .	Mathematical Practice 8 When students look for and express regularity in repeated reasoning, they . . .
Begin with facts; using other facts, the logic leads to factual conclusions as in a geometric or algebraic proof	Begin by exploring and recognizing patterns
Use a system of patterns and mathematical properties	Use identified patterns to make conjectures
Identify and use structure to make sense of the mathematics	Make a generalized statement from a conjecture
Know that the conclusions are correct and can prove them	Think the generalized statement is correct but have not established the validity through a proof
Think deductively	Think inductively
Solve problems such as $3 + 5 + 7 + 5 = ?$	Solve problems such as Given: 3, 6, 9, 12 . . . What is the 10th term?

Students make use of structure (Mathematical Practice 7) in the last example in the chart as they use the associative and commutative properties to group the values to make tens. Students express regularity in repeated reasoning (Mathematical Practice 8) in the last example in the chart as they recognize that the number of each term multiplied by 3 produces the value of the term. Thus, the tenth term must be 30. Generalizing the pattern helps students predict any term in the pattern.

With the distinction between Mathematical Practice 7 and Mathematical Practice 8 clarified, the rest of this chapter is devoted to Mathematical Practice 8, "Look for and express regularity in repeated reasoning."

Figure 8.2 provides a sample task showing Mathematical Practice 8.

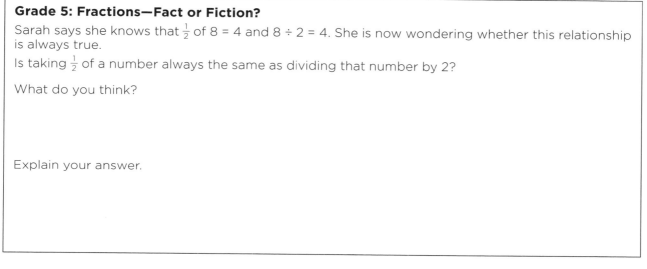

Grade 5: Fractions—Fact or Fiction?

Sarah says she knows that $\frac{1}{2}$ of 8 = 4 and 8 ÷ 2 = 4. She is now wondering whether this relationship is always true.

Is taking $\frac{1}{2}$ of a number always the same as dividing that number by 2?

What do you think?

Explain your answer.

Figure 8.2: Fractions—fact or fiction?

*Visit **go.solution-tree.com/MathematicsatWork** for a free reproducible version of this figure.*

Students in Ms. Mulcauhey's fifth-grade class are working on multiplying fractions. They have used fraction strips, fraction circles, and area models. With this task, she is presenting an opportunity for students to connect the relationship between multiplication and division.

Students work on the question from figure 8.2 independently at first. Ms. Mulcauhey then asks them to join their teams and share their thinking. Following are some of her students' thoughts.

- "I made a list, and it looks like taking half of a number is the same as dividing by 2."

- "I'm wondering if this is true when the number is a fraction. Is $\frac{1}{2}$ of $\frac{3}{4}$ the same as $\frac{3}{4}$ ÷ 2?"

One student provides the following response. "She is correct. When you have an even number you can divide it by two or find half of it, but not if it is an odd number."

This student agrees with the statement presented in the question, but qualifies her answer by limiting the situation to only even values.

Another student provides the following response, showing her reasoning and providing specific examples (see figure 8.3). Note the reference to the number trick and the up arrows (that indicate the denominator going into the numerator) that appear in the example.

Does this student understand *why* the procedure works, or has she simply memorized the steps?

$$1. \frac{1}{2} \times 9 = \frac{9}{2} = 4\frac{1}{2} = 9 \div 2 = \frac{9}{2}\uparrow = 4\frac{1}{2}$$

$$2. \frac{1}{2} \times 6 = \frac{6}{2}\uparrow = 3 = 6 \div 2 = \frac{6}{2}\uparrow = 3$$

Yes, it is true that the answer will be the same because I multiplied with an even and an odd number. You can keep, change, flip trick with the division. First you would have to keep the first number or fraction alone. Then you would change the symbol to the opposite. Next you would have to flip the fraction into a whole number or a whole number into a fraction.

Figure 8.3: Fractions—fact or fiction? student response.

These students have noticed the repetition and are now looking for generalizations. They are making predictions and reflecting on whether their ideas are making sense. Students are engaged in Mathematical Practice 8.

Table 8.2 shows more examples for what students do during a lesson to demonstrate evidence of learning Mathematical Practice 8 and what actions you can take to develop this critical thinking in students.

Table 8.2: Student Evidence and Teacher Actions for Mathematical Practice 8

	Student Evidence of Learning the Practice	Teacher Actions to Engage Students

Look for and express regularity in repeated reasoning.	Students:
	
	• Notice repetition and patterns in counting and calculations
	• Organize their repeated trials to help identify existing patterns
	• Label all quantities being used in the problem to make sense of the problem and recognize repeated reasoning
	• Look for generalizations and identify shortcuts based on the repeated patterns
	• Compare generalizations looking for similarities and differences
	• Continually ask, "Does this make sense?"
		Teachers:
		
		• Provide rich and varied tasks for students to explore
		• Provide ample time for students to engage in the problem-solving process, including ample opportunities for students to share their thinking and justify their reasoning
		• Suggest ways to organize students' work to make patterns more visible
		• Ask questions such as:
		  • "What do you notice . . ."
		  • "How can you describe this pattern?"
		  • "What would happen if . . ."
		  • "Have we seen this before?"
		  • "What predictions or generalizations can you make based on this pattern?"
		  • "How might your work on the first problem help you now?"

*Visit **go.solution-tree.com/MathematicsatWork** for a free reproducible version of this table.*

Understand *Why*

Mathematics can be thought of as the study of patterns. Mathematics offers students opportunities to recognize patterns and make predictions based on their discovery of patterns. As Arthur Hyde (2006) says:

> One of the greatest gifts we can give our students is to help them develop their innate capability to *infer patterns and then use inferences to predict.* Helping children become more sophisticated in their inferences and their predictions is a major job for the teacher. Whether the patterns are numerical, visual, auditory, or otherwise, the cognitive processes are very similar and they are greatly facilitated by language. (pp. 114, 115)

As children engage in looking for and identifying patterns, they are making sense of the mathematics. They are constructing their knowledge and by doing so, they can then extend their understandings to new areas. But, what does it mean to *understand?*

David Perkins (1998), professor at Harvard University Graduate School of Education, defines understanding as "the ability to think and act flexibly with what one knows." Notice that this definition involves actively using and applying knowledge. Perkins does not mention rote learning or memorization as a characteristic of understanding.

Grant Wiggins (2014) writes, "Mathematical understanding is the ability to justify, in a way appropriate to the student's mathematical maturity, why a particular mathematical statement is true or where a mathematical rule comes from." The Common Core State Standards use the term *understand*, but the reader is left to infer the meaning of what it means to understand. Grant Wiggins (2014) further writes in this same blog entry that "understanding requires focused *inferential work*. Being helped to generalize from one's specific knowledge is key to genuine understanding."

Kilpatrick and colleagues (2001) define conceptual understanding as "an integrated and functional grasp of mathematical ideas" (p. 118). Additionally, they state, "Knowledge that has been learned with understanding provides the basis for generating new knowledge and for solving new and unfamiliar problems" (p. 119). Thus, the process of looking for and identifying patterns, extending those patterns, suggesting generalizations, and perhaps generating new knowledge builds students' mathematical understanding. As students engage in this work, they are immersed in Mathematical Practice 8, "Look for and express regularity in repeated reasoning."

This chapter focuses on tasks involving two types of patterns—(1) repeating patterns and (2) growing patterns. *Repeating patterns* are introduced early in a child's development and are experienced continuously thereafter. The representations of repeating patterns vary greatly from sounds, objects, and numerical patterns to algebraic and geometric patterns. It is important that these patterns are presented illustrating the repetition of the core of the pattern. Van de Walle and Lovin (2006) explain, "The *core* of a repeating pattern is the shortest string of elements that repeats" (p. 291). For example, given the core of a repeating pattern is ABA, stating the pattern as ABAA is ambiguous. Two repetitions need to be presented—ABA ABA—to ensure clarity.

Growing patterns are the second type of patterns that students can explore. Growing patterns or sequences "consist of a series of separate steps, with each new step related to the previous one according to the pattern" (Van de Walle & Lovin, 2006, p. 293). Growing patterns reveal a change from one term to the next. Remind students that in spite of their name, growing patterns can also decrease from one term to the next (NCTM, 2001). Teachers can provide opportunities for students to record patterns in various forms such as using objects, making a table, and creating a graph. Through the use of multiple representations, students can explore to discover relationships and make generalizations about those relationships. (See more about repeating and growing patterns on pages 215–222.)

Looking for the patterns' relationships and stating generalizations or conjectures about them demonstrate Mathematical Practice 8, which is expected of mathematically proficient students. According to Mark Driscoll (1999):

> The habit of thinking that causes one to look beyond a perceived pattern to wonder what "always works" for the pattern's mathematical rule is a feature of a broader capacity of mathematical thinking; the recognition, expression, and manipulation of generalities, commonly called *generalization*. (p. 91)

Young students can participate in this generalization process. Consider the following examples.

Given: 12 + 3 – 3 = ?
4 + 8 – 4 = ?

Students can explain that adding a value and then subtracting that same value does not affect the remaining value regardless of where the numbers are written in an expression involving a string of addition and subtraction symbols. Creating and reflecting on the validity of generalizations are critical components that deepen students' mathematical understanding. When proven, this understanding will eventually lead to the zero property of addition, a structure that students can use throughout mathematics (Mathematical Practice 7).

As students explore patterns, multiple representations help them. As Hyde (2006) explains, "Multiple representations also may provide deeper, more elaborate understandings of the underlying mathematics, and fresh, new insights into the problem" (p. 87). Students can build the patterns using a variety of materials such as Unifix cubes, cotton balls, and so on, and they can draw diagrams. Students can also create a table of values as they search for relationships between the number of each term in the sequence and the resulting quantity in the term.

When students learn the habits of mind to look for and express regularity in repeated reasoning (Mathematical Practice 8), they begin to notice patterns on their own and generalize those patterns verbally and in writing.

Strategies for *How*

When teaching students to look for repeated reasoning and express the regularity of that reasoning, it is critical to use intentional tasks. For each strategy, notice the specific patterns students must identify or explore.

Exploring Repetition

Young students notice and enjoy repetition. Ever witness how easily babies get the adults around them to repeat a game over and over again? Young children can get adults to play peekaboo, pick up a dropped toy from a high chair, and make a special funny face repeatedly. The responses the adult and child give and receive encourage this playful repetition. When asked to select a bedtime story, children have their favorites. Night after night, children select the same story. Why? They enjoy the repetition and knowing what comes next.

Patterns exist everywhere. Encourage students to notice patterns all around them. They can see patterns in their own movement, in their clothes, in music, in nature, in the routines in their classrooms, and in their own routines at home. Recognizing, describing, and extending patterns encapsulate Mathematical Practice 8.

Preschool and kindergarten students enjoy being part of a pattern. Here are some ways to have young students actively participate.

Jump, Stomp, Clap

This strategy is one way to have students experience patterns. Ask students to join you in an open area. Suggest that they are going to be a part of the pattern and they will practice first. Demonstrate the pattern by jumping once with both feet, stomping with one foot once, and clapping your hands one time. Repeat. Invite students to join you. Repeat the pattern several times.

Next, tell students to determine what comes next in the pattern. Demonstrate the same pattern, but then stop. Ask, "What do we do next?" Repeat several times, stopping at various points in the pattern.

Have students form a large circle. Students will demonstrate one action in the pattern on their turn. Tell students to pay attention, as each student will show the next action in the pattern.

Ask students to describe the pattern. How many actions did they have in the pattern? In which order did the actions appear? Is the pattern different if we start with a stomp rather than a jump? With practice and support, you can begin to generalize the pattern with an alphabet structure, such as the A-B-C pattern.

Repeat this activity using other actions and pattern sequences.

Evens Up

Evens Up is an activity that helps grade 2 students recognize even and odd numbers. As students sit in a circle, have them count off 1, 2, 1, 2 until everyone has a number. Tell students that they are going to repeat the pattern, but this time, if they say "2," they will stand up. After everyone in the class has participated, ask students what they notice. Look for students to mention that every other student is standing. Then, ask the standing students to remain standing. Tell students that they are now going to count by ones. Have one student who is sitting begin counting. The count continues around the circle until every student has spoken. Ask students what they notice. If students do not mention that all even numbers are standing, ask them to repeat counting by ones, and give each student a card with the appropriate number on it. Then, ask what they notice. Extend this work to ask students if they counted to 52, would a person with that number be sitting or standing? How do they know? Making predictions and expressing how the pattern repeats incorporate thinking strategies students require for Mathematical Practice 8.

Hundreds Chart

Exploring patterns using a hundreds chart provides bountiful opportunities for students to look for and express regularity in repeated reasoning. With numbers on a number line, the repeated patterns may not be as readily apparent to young learners. Consider the hundreds chart in figure 8.4.

This hundreds chart starts at the upper left and proceeds along to the right. The numbers get larger as they continue across the chart and as they proceed down the chart. Unfortunately, this may be counterintuitive for young students as they tend to connect up with larger values and down with smaller values. Just because lines of text are read down the page does not mean that number sense should use that same format. Consider using more precise language such as "greater than" and "less than" in these situations. For some students, seeing this hundreds chart built up from the bottom connects the greater numbers above the smaller ones as in figure 8.5. Notice that this chart starts with zero to emphasize the same tens digit begins in each row and is still read left to right while reading up.

1	2	3	4	5	6	7	8	9	10
11	12	13	14	15	16	17	18	19	20
21	22	23	24	25	26	27	28	29	30
31	32	33	34	35	36	37	38	39	40
41	42	43	44	45	46	47	48	49	50
51	52	53	54	55	56	57	58	59	60
61	62	63	64	65	66	67	68	69	70
71	72	73	74	75	76	77	78	79	80
81	82	83	84	85	86	87	88	89	90
91	92	93	94	95	96	97	98	99	100

Figure 8.4: Hundreds chart.

100	101	102	103	104	105	106	107	108	109
90	91	92	93	94	95	96	97	98	99
80	81	82	83	84	85	86	87	88	89
70	71	72	73	74	75	76	77	78	79
60	61	62	63	64	65	66	67	68	69
50	51	52	53	54	55	56	57	58	59
40	41	42	43	44	45	46	47	48	49
30	31	32	33	34	35	36	37	38	39
20	21	22	23	24	25	26	27	28	29
10	11	12	13	14	15	16	17	18	19
0	1	2	3	4	5	6	7	8	9

Source: Didax, 2015. Used with permission.

Figure 8.5: Hundreds chart with an alternate organization.

Presenting the chart in this manner helps students see the repeated pattern in the numbers. To emphasize the repeated patterns, arrange strips of values in each row and place the strips end to end as in a number line. Then, students can build the hundreds chart by beginning with the 0–9 strip and place the 10–19 on top of the first strip. This process can repeat as students are counting together to reach 100 and then on to 120. The next row is placed on top of the previous one. This action also increases students' awareness of completing one group of ten and moving onto the next. Students might also notice that the entire pattern repeats after 100. Students identify the counting sequence in the row of values. Ask students to predict what values will be in the next row of numbers. Other examples of patterns include skip counting and recognizing changes in the tens and ones digits when adding and subtracting by 1 or 10.

As students identify patterns, they extend their thinking by making generalizations about the pattern. There are a vast number of resources featuring explorations on hundreds charts for students. T-Time and X-It are two activities appropriate for students who are developing skills in adding two-digit numbers.

T-Time

In T-Time, students select a square on the hundreds grid that is *not* a value on the border, such as 14 or 87 as compared to 3 or 98, which occur on the upper and lower borders. Students then write down the value directly above and below the value in their chosen square. They also write down the value that is on the left and the value on the right of the first square.

Students then explore relationships. What do they notice about the sums of the two pairs of numbers? Is this pattern consistent with any values that they might select? What are they thinking as to why this relationship occurs? For example, suppose a student chooses 16. They then explore the relationship between the sum of 6 and 26 as compared to the sum of 15 and 17. Have students explain the patterns using their knowledge of tens and ones.

X-It

X-It is similar to T-Time, but students choose the diagonals of the selected number. Students select a square that is not on the border, as in T-Time. Students can imagine that their selected square is in the middle of a three-by-three square. They then identify the numbers located in the upper-left corner and the lower-right corner of their three-by-three square. Students also write down the values in the upper-right corner and lower-left corner.

What do they notice about the sum of these values? Does this occur anywhere on the chart? If the chart went beyond 100, would the pattern exist? Suppose a student selects 34 and will then explore the sum of 45 + 23 as compared to 43 + 25. What will that student notice? See figure 8.6 for the student's sample grid.

A monthly calendar has similar values. The values are simply restricted to smaller numbers in this context.

50	51	52	53	54	55	56	57	58	59
40	41	42	43	44	45	46	47	48	49
30	31	32	33	34	35	36	37	38	39
20	21	22	23	24	25	26	27	28	29
10	11	12	13	14	15	16	17	18	19
0	1	2	3	4	5	6	7	8	9

Figure 8.6: Sample X-It exploration grid.

Discovering Patterns

Discovering patterns can occur in just about every lesson when teachers *intentionally plan* for the opportunity. Students can always look for relationships and patterns. In kindergarten, students can explore the meaning of adding one more unit as they use the number line. In first grade, they can explore what happens when they add 10 to a quantity. As they recognize the patterns and search for repetition in the pattern, students engage in Mathematical Practice 8. When they begin to verify why the pattern is true based on facts that they already know, students are thinking in terms of Mathematical Practice 7, "Look for and make use of structure." The following tasks provide a few more examples of how to implement Mathematical Practice 8, "Look for and express regularity in repeated reasoning."

In the primary grades, you may give students a task such as the one in the following example.

> Charlie has 13 yummy gummy bears. He wants to share them with his friend, Mario. How many gummy bears will each person have?

Provide students with a similar context. (You may want to increase or decrease this quantity of candy depending on the needs of your learners.) Have bear counters, chips, and small strips of paper for students to select from as they begin working on this problem. The tool they select does inform teachers as to the student's level of mathematical thinking in terms of their ability to think abstractly. Students who draw pictures of the situation are beginning to work more abstractly than students who want to work solely with bear counters. As students work, look for some partners to ignore the fact that one person has less than the other. Also, look for students to change their tool so they can cut the last piece of candy in half or represent half in their drawing.

To intentionally work with students to recognize repeated reasoning and discover a pattern, ask students questions such as, "Did both boys get the same amount?" and "Would they if they started with a different number of gummy bears?" Create a chart like figure 8.7 (page 214) for students to use as a reference.

Two Students Sharing	
Number of Candies	**Equal Shares?**
13	no
7	no
8	yes

Figure 8.7: Student shares pattern.

Students can now explore the relationship between the number of candies and whether two students will receive an equal amount when they share, provided no piece of candy is cut in half. Encourage students to reflect on whether their answers make sense as they work. What do they notice? What repeats? Can one entry in the chart help with the next entry in the chart?

In the intermediate grades, the task may change, as in the following example.

> Felicia has one large sandwich. She wants to share this sandwich with three of her friends. How much of the whole sandwich will the four children get to eat?

Look for some students to represent the sandwich by drawing it or using Unifix or linking cubes. After students have had a chance to explore this context, create a chart like figure 8.8 for them to record their findings and discover patterns related to other combinations regarding the number of students sharing a number of sandwiches.

Number of Students	Items Shared	Amount in Each Share
4	1 sandwich	$\frac{1}{4}$ sandwich
5	1 sandwich	
7	1 sandwich	
7	2 sandwiches	
8	1 sandwich	
8	5 sandwiches	

Figure 8.8: Student recording chart.

As students collaborate on these situations, ask them to predict the amount in each share before beginning their work. Just as students are asked to make predictions as they read a story, predictions based on patterns help to focus students' attention and extend their mathematical understanding. As they proceed, consider asking them the following questions.

- "Does your answer make sense?"

- "What is repeating?"

- "What pattern are you noticing?"

- "Why might this be true?"

Students can also complete the following sentence frame.

"Knowing _____, can you predict the answer to _____?"

These explorations can lead students to ask "What if?" questions such as "What if the number of sandwiches is greater than the number of students?" and "What if there is only part of one whole item to share?"

Some lessons naturally include discovery of patterns. In those lessons that do not naturally lend themselves to patterns, challenge yourself to think about a pattern students can discover, and determine the best way to have them implement this critical-thinking skill.

Identifying Patterns

Hyde (2006) states that "one of the greatest gifts we can give our students is to help them develop their innate capability to *infer patterns and then use inferences to predict*" (p. 114). Detecting, identifying, extending, and generalizing patterns enable students to build their understanding of mathematical relationships. Patterns can be expressed in words and symbols, and the same pattern may be seen in a variety of forms. Discovering patterns and identifying patterns are closely related. Students discover patterns when they explore solutions to problems or wonder about number or shape relationships. After many explorations, students might discover there is a general pattern. As students discover patterns, they apply the mathematics they understand. They are able to extend the pattern and perhaps make conjectures about the pattern. Students work to identify a pattern, and then they state the relationship within it. For the purpose of this discussion, two fundamental patterns, repeating and growing patterns, will be featured.

Repeating Patterns

In first grade, you might give students a geometric pattern like the one in figure 8.9.

Figure 8.9: Repeating geometric pattern example 1.

Next, ask students, "What shapes come next if this pattern continues?" Young students should work with actual shapes. Older students can extend the pattern and determine what the fifteenth shape will be. Students should provide a prediction for their answer and explain their thinking. Students then check their prediction.

Also, ask young students to compare patterns. Consider showing students the pattern in figure 8.10 and asking them if the pattern is the same as that in figure 8.9. Students who identify patterns using an alphabet structure understand that both of these patterns represent an ABB pattern. Even though geometric shapes are different, the patterns' structure are the same.

Figure 8.10: Repeating geometric pattern example 2.

A more challenging question for students to explore is determining how many shapes will be in the pattern when the ninth triangle is drawn. Students may recognize that the pattern builds on multiples of 3, and they might predict that there will be twenty-seven shapes in the pattern. As the triangle begins the pattern, students should not include the two circles after the ninth triangle. This means there will be twenty-five shapes when the ninth triangle is drawn.

Patterns are typically in a line. Consider presenting patterns in a grid like the one in figure 8.11.

This ABC pattern presents many areas for students to explore. What do they notice about the patterns in the grid? Do they see the repeating shapes in the diagonals? Two circles appear in the second and fifth rows. When will this occur again?

The preceding examples feature repeating patterns using shapes. Students begin working with shapes and colors and then explore repeating patterns involving numbers. An example of a repeating pattern using values is 1, 2, 3, 4, 1, 2, 3, 4, 1, 2, ____, ____.

After identifying the core of the pattern, the values that repeat, you can ask students to predict what numbers should go in the blanks. They can also predict what value would be in the twentieth position in the pattern.

Growing Patterns

Although young students focus more on repeating patterns, they should also experience growing patterns. A growing pattern is one that changes from one term to the next, each increasing in size, using a predictable rule.

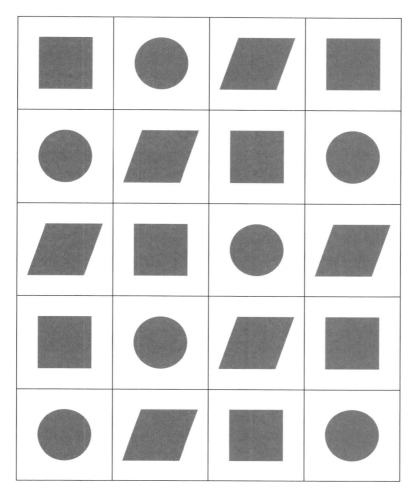

Figure 8.11: Grid pattern example.

*Visit **go.solution-tree.com/MathematicsatWork** for a free reproducible version of this figure.*

Consider the numerical growing patterns in the following example. Each one involves addition or multiplication patterns from one term to the next.

> 1, 3, 5, 7, 9 . . . Each term is 2 more than the previous term.
> 2, 4, 8, 16, 32 . . . Each term is 2 times the previous term.
> 2, 3, 5, 8, 12 . . . The difference between the terms is +1, +2, +3, and so on.

The next two sections will explore growing patterns that are geometric and numerical.

Growing Geometric Patterns

Some growing patterns use geometric shapes. Figure 8.12 illustrates such patterns. Students can predict the next shapes in each pattern, and then they can determine whether their predictions are correct.

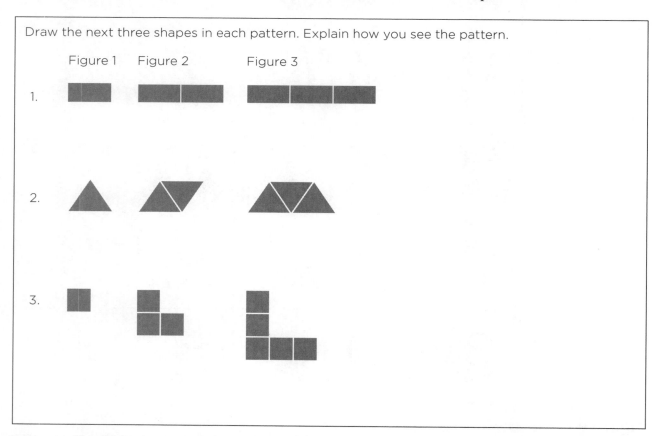

Figure 8.12: Growing shape patterns.

*Visit **go.solution-tree.com/MathematicsatWork** for a free reproducible version of this figure.*

As students gain experience with growing patterns, they can number each step in the series. They then can use the step number (or figure number) to identify quantities that occur later in the series. At this point, students make predictions about the number of items in that later step, as it may contain too many items to actually build. It is helpful for students to create a chart or table to represent these patterns and to identify any possible relationships between the number of the terms and the items in each term. For example, figure 8.13 refers to the third question in figure 8.12 and shows the appropriate chart that students might create to predict the number of squares later in the pattern. This example merges the idea of a growing geometric and numerical pattern.

Figure	Number of Squares
1	1
2	3
3	5
4	?
8	?

Figure 8.13: Using a chart to make sense of a pattern.

Growing Numerical Patterns

Providing students with a context helps them reason about numerical patterns. The following sample task presents a context for lower elementary students to explore a growing numerical pattern.

> Johnny had a dream. In his dream, every day he received one more dollar than he received the day before. On Monday, Johnny got $1. On Tuesday, Johnny received $2. How many dollars will Johnny get on the fifth day if his dream comes true? Show how you know.

Students may build the pattern for this task using play money or chips or by creating a table of values. Teachers can ask students to describe how the money increases from one day to the next. Asking students to determine the total amount of money that Johnny has received by the fifth day can extend this task. Some students mistakenly think that this task represents a repeating pattern (1, 2, 1, 2). By describing the pattern in terms of getting one more dollar each day, students can then work with money or chips to build the growing pattern.

Older students can use the same task as they think about this pattern considering two values. See figure 8.14. Students can keep track of the dollar amount increasing by one each day as well as how the total amount of money changes from one day to the next.

Day	1	2	3	4	5
Dollars Received	1	2	3	4	5
Total Dollars	1	3	6	10	15

Figure 8.14: Money in Johnny's dream.

Ask students to identify the pattern in terms of the total number of dollars that Johnny might receive each day. Students will easily recognize that the dollars received each day increase by one. When asked to recognize a pattern with the total dollars, students might (1) recognize they are adding +2, +3, +4, and so on between total dollar terms or (2) add the day number, or dollars received, to the previous total dollar amount (for example, 1 + 2 = 3 and 3 + 3 = 6, and so on). Some students will extend the existing chart as others choose to work with play money or chips to represent the dollars. They can also predict how much money Johnny might have after eight days or ten days. Teachers can use tasks like this one in several grade levels and even repeat them with the same students from one grade to the next.

Figure 8.15 presents another numerical pattern using a grid. In this What Do You Think? activity, students consider the relationship between the values in the rows and columns, and demonstrate their mathematical thinking and their level of understanding as they identify the patterns that they see in the grid.

Row	Column					
---	A	B	C	D	E	F
1	1	2	3	4	5	6
2	7	8	9	10	11	12
3	13	14	15	16	17	18
4	19	20	21	22	23	24
5						
6						
7						
8						

This pattern continues forever. What number appears in column C, row 10?

In what column and row does 72 appear? Justify your reasoning.

Figure 8.15: What Do You Think? example.

*Visit **go.solution-tree.com/MathematicsatWork** for a free reproducible version of this figure.*

Primary students might approach What Do You Think? by counting on. They might recognize that each column increases by 6 from the original value. Intermediate students might approach the second question by thinking about multiples of 6, for example, 6 × 12 is 72, so 72 will appear in column F, row 12.

Students should have opportunities to look for patterns before you ask specific questions. Teachers can collect what students notice about the grid as well as what questions they may have. Those questions may lead to further explorations and generalizations about the patterns.

An In and Out chart provides another structure in which students can look for and express regularity in repeated reasoning. Consider the In and Out chart in figure 8.16.

In	Out
0	4
1	5
2	6
3	7
4	?

Figure 8.16: In and Out chart.

In figure 8.16, students look for the regularity in this pattern to determine that each value in the In column increases by one. The values in the Out column also increase by one from one term to the next. Also, each Out value can be seen as four greater than its corresponding In value. Students state that the 8 replaces the question mark and justify their thinking by orally presenting the rule they found governing the pattern.

Students should see the In and Out charts both horizontally and vertically. Figure 8.17 shows a horizontal example.

In	0	1	2	3	4
Out	4	5	6	7	?

Figure 8.17: Horizontal In and Out chart.

Students can create an In and Out chart as well as use it to identify a pattern or determine a missing value in a pattern. In kindergarten and first grade, you can use number tracks to create an In and Out chart. See figure 8.18 (page 222) for an example. In this figure, the In and Out, or Starts and Lands, tool provides students with a structure to determine the pattern explained in words and numbers and, in this case, extend the pattern. You can provide students with opportunities to discuss their reasoning with others to deepen their understanding.

| 1 | 2 | 3 | 4 | 5 | 6 | 7 | 8 | 9 | 10 | 11 | 12 |

My little brother stands on the number track at 2. When I say "Go," he jumps and lands on 4. When I say "Go" the next time, my brother jumps and lands on 6. Where will he land the next time I say "Go"?

Record your work in the table.

| **Starts** | 2 | 4 | 6 | ? |
| **Lands** | 4 | 6 | ? | ? |

When I say "Go" five times, where will my brother be on the number track?

Figure 8.18: Number track task.

Visit **go.solution-tree.com/MathematicsatWork** *for a free reproducible version of this figure.*

Representing Patterns

Once students recognize a pattern, how can they explain or represent the pattern to make it clear for others? Students can use words, pictures, tables, numbers, and even formulas to represent patterns they notice. Consider the fifth-grade task in figure 8.19. Working in groups, students conclude that in the pattern shown, each new term is two greater than the one before it. When asked how many circles would be in the fifth figure, students easily respond that the fifth figure would have nine circles. The teacher then asks about the tenth figure. Some students think there would be eighteen circles in the tenth figure (rather than the correct answer of 19), as that would be the amount that doubled the circles in the fifth figure. Teams then begin to work on a chart to help determine the number of circles in the tenth figure.

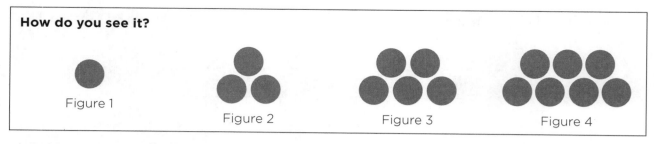

How do you see it?

Figure 1   Figure 2   Figure 3   Figure 4

Figure 8.19: How Do You See It? grade 5 task.

Extending a typical In and Out chart to include a middle column labeled *What I See* encourages students to articulate how they are viewing a pattern. *Figure Number* replaces *In*, and *Total Circles* replaces *Out*. See one fifth grader's chart in figure 8.20.

Figure Number	What I See	Total Circles
1	1 circle	1
2	1 circle + 2	3
3	1 circle + 4	5
4	1 circle + 6	7
5	1 circle + 8	9

Figure 8.20: Marcus's representation of the pattern.

Another student, Adisa, sees the configuration as a rectangle with one circle missing (see his chart in figure 8.21).

Figure Number	What I See	Total Circles
1	1 circle	1
2	4 circles, take away 1	3
3	6 circles, take away 1	5
4	8 circles, take away 1	7
5	10 circles, take away 1	9

Figure 8.21: Adisa's representation of the pattern.

After working awhile, students come back together to share their thinking about the number of circles that would be in the tenth pattern. Most of the class builds the pattern by drawing all of the terms. Others speak about the relationship between the term number and the quantity of circles.

Lucia: I saw that as the figure number increased by 1, the total got bigger by 2.

Roberto: Are we always going to have to build the pattern?

Adisa then shares his chart with the class. Together, the class realizes that the number of circles that Adisa saw in his head was double the figure number. They then conclude that the total was one less than two times the figure number.

These students used a diagram, chart, and then a verbal description of the pattern. With support from their teacher, they described the pattern symbolically as *Total = 2n − 1*, with *n* representing the figure number. It is important to explicitly make connections between these representations with students. Ask, "Why did we include − 1 in the equation?" "How does the chart reflect the need to double the figure number?"

Students' varying statements demonstrate different ways of thinking about these patterns. Lucia saw a recursive pattern as she used the previous term to determine the next term. Adisa identified an explicit pattern. He doubled the figure number and then subtracted one circle in each case. Students will do more with these patterns in middle school and high school. For now, students should begin to recognize that if they need to determine the one hundredth term in an explicit pattern, an equation or formula is more efficient and effective. Equations and formulas are the generalizations that Mathematical Practice 8 emphasizes.

Playing Games With Patterns

There are many games that help students recognize and use patterns. Commercial games such as Qwirkle, checkers, dominoes, and Simon all encourage students to focus on patterns and develop specific strategies. The following two games, Poison and tic-tac-toe, are easy to implement in the classroom.

Poison

This game uses thirteen chips, twelve of one color and one of another color. The single color represents poison. Two players place all the chips on the playing surface. Figure 8.22 shows the chips in a line, but this does not have to be the case. Chips can be in a pattern or scattered around on the playing surface. Each player, in turn, may take one or two chips away from the group. The object of the game is to force the opponent to take the last chip, as it represents the poison.

Figure 8.22: Poison game.

When playing Poison, consider pairing students with a partner. Instruct students to develop a winning strategy. Encourage partners to play several practice rounds and then discuss their strategy. Listen for students to make conjectures such as "the student who goes first wins" or "if we always think about groups of 3, we can win." After students have recorded their winning strategy on paper, have students use their strategy playing against another group. After playing a few rounds against another team, do they need to reconsider their strategy?

A winning strategy for Poison involves recognizing that you want to make certain the opponent's turn occurs when there are only four chips remaining, three regular chips and the poison chip. As the opponent can only remove one or two chips, two turns remain. To accomplish this, some students will realize on their turn they want to make sure a multiple of three chips has been removed (3, 6, 9) which means the opponent will eventually have four chips remaining.

If player 1 takes two blue chips, one blue chip remains along with the poison. Then, player 2 takes the one blue chip forcing player A to take the poison. If player 1 takes one blue chip, then player 2 takes two blue chips, leaving player 1 with the poison chip.

Tic-Tac-Toe

Tic-tac-toe has never lost its appeal for students in the elementary grades. To implement this game in a center, it is important that students consider a winning strategy. Ask students questions similar to the following.

- "What is a good first move?"
- "Can we predict what our opponent might do in response?"
- "Is there a pattern in moves that tends to lead to a tie game?"

Students can play a few rounds and then try out their strategy as they play other partners.

Winning strategies involve both offensive and defensive approaches. Many students select the center square as the opening move, but that does not guarantee success. Students try to create a situation in which they have two ways to win, and the opponent can only block one of those ways.

Another version of tic-tac-toe uses a board with a mathematics problem in each square. Small pieces of paper cover each problem, and on each turn, the student must answer the problem before placing an *X* or *O* on the square. Of course, the opponent verifies the student's answer is correct.

Lesson Example for Mathematical Practice 8: Problems With Patterns

The lesson plan in figure 8.23 (pages 226–228) focuses on using patterns to solve a problem. Fifth-grade students will recognize patterns and extend the patterns to new situations. Using a context of baking cookies, they will generate two numerical patterns and draw conclusions based on resulting tables and graphs. First, students will review their understanding of plotting points on a coordinate plane and review the attributes of geometric shapes as they answer questions using the first quadrant on a coordinate plane. Although Mathematical Practice 8 is the focus of the lesson, students will also provide explanations of their reasoning (Mathematical Practice 2). A commentary follows the lesson providing more information related to the rationale and importance of each lesson component. Figures 8.24 and 8.25 (pages 228–231) support the lesson's tasks.

Unit: Operations and Algebraic Thinking (5.OA.3 and 5.G.3)

Date: May 15

Lesson: Problems With Patterns

Learning objective: As a result of class today, students will create tables and graphs representing a real-world problem and draw conclusions based on the data.

Essential Standard for Mathematical Practice: As a result of class today, students will be able to demonstrate greater proficiency in which Standard for Mathematical Practice?

Mathematical Practice 8: "Look for and express regularity in repeated reasoning."

- Students will look for patterns and apply them to answer questions in the context of baking cookies.
- Students will draw conclusions based on patterns.
- Students will explain their thinking and reflect on whether their answer makes sense.

Formative assessment process: How will students be expected to demonstrate mastery of the learning objective during in-class checks for understanding teacher feedback, and student action on that feedback?

- Students will work with partners sharing their conjectures and drawing conclusions.
- Students will supply feedback to each other and reflect on whether their answers make sense.
- Teachers will monitor students' progress and provide feedback by asking assessing and advancing questions.

Probing Questions for Differentiation on Mathematical Tasks

Assessing Questions	Advancing Questions
(Create questions to scaffold instruction for students who are stuck during the lesson or the lesson tasks.)	(Create questions to further learning for students who are ready to advance beyond the learning standard.)
<ul><li>What do we know, and what do we want to find out?</li><li>Could you make a table of values to help guide your thinking?</li><li>How might two tables help you compare the two baking times?</li><li>What conclusion can you make by looking at the graph?</li></ul>	<ul><li>What are you wondering about?</li><li>What other times will the two batches of cookies be ready together?</li><li>If you connected the points on the graph for one type of cookie, what do you think you might see?</li><li>Is this occurring for both types of cookies? Why do you think this is happening?</li><li>Write a question that involves patterns. Be sure to remind students to include a table and a graph.</li></ul>

Tasks (Tasks can vary from lesson to lesson.)	What Will the Teacher Be Doing? (How will the teacher present and then monitor student response to the task?)	What Will Students Be Doing? (How will students be actively engaged in each part of the lesson?)
Beginning-of-Class Routine How does the warm-up activity connect to students' prior knowledge, or how is it based on analysis of homework?	Introduce the lesson's focus on patterns. Show students the first quadrant of a coordinate plane. Ask them to locate (3,5). Ask, "Is (3,5) the same as (5,3)? Explain your thinking." Review the meaning of attributes of basic geometric shapes by asking, "What attributes do squares and rectangles have in common?" Make a list of the attributes that students suggest. Look for students to suggest four right angles, opposite sides are parallel, at least two sides are congruent, and adjacent sides form right angles. Tell students that they will use these attributes as they answer questions during task 1, What's the Point?	Students think about the location of (3,5). After they identify the location, they then consider whether (3,5) is the same as (5,3). Students share their response with a partner. Students think independently and then share their ideas with the class.
Task 1 How will students be engaged in understanding the learning objective? (See figure 8.24.)	Monitor the room to listen to students' conversations as they discuss their responses to the questions in task 1. Consider writing down what you hear them say. Listen for mathematical ideas including vocabulary justifying the rectangle and squares drawn on the coordinate planes (for example, opposite sides are equal or congruent) as well as evidence of collaborative skills. Consider sharing these statements with the class as a part of the closure.	Students again think independently for a few minutes, and then they will join their partners to compare their answers to questions in task 1.
Task 2 How will the task develop student sense making and reasoning? (See figure 8.25.)	Provide students with a copy of The Big Cookie Bake activity sheet (figure 8.25, pages 230–231). Have students read the problem, and then ask any clarifying questions. Be prepared to help some students make sense of the tables so they can complete them accurately. Start with 0 batches, and ask students, "How many minutes does it take to bake 0 batches?" (Answer: 0 minutes)	Students read the task and have time to ask any clarifying questions. They first work independently. On signal, they confer with their partners and share their understanding.

continued →

Closure		
How will student questions and reflections be elicited in the summary of the lesson? How will students' understanding of the learning objective be determined?	Call on student partners to share with the class what they noticed and what they are still wondering about. What information did they get from the table? What information did they get from the graph? Ask students to reflect on why mathematicians consider patterns to be powerful. What do they think? Consider sharing the statements that you heard students making during task 1 and task 2. Feature any mathematical relationships as well as evidence of students collaborating effectively.	Partners share their findings with the class. Exit ticket questions: As a result of the lesson today, I know that _____. A question I have is _____.

Source: Template adapted from Kanold, 2012c. Used with permission.

Figure 8.23: Grade 5 lesson-planning tool for Mathematical Practice 8.

*Visit **go.solution-tree.com/MathematicsatWork** for a free reproducible version of this figure.*

Task 1: What's the Point?   **Name:** _____

1. Write the coordinates for each point on the plane.

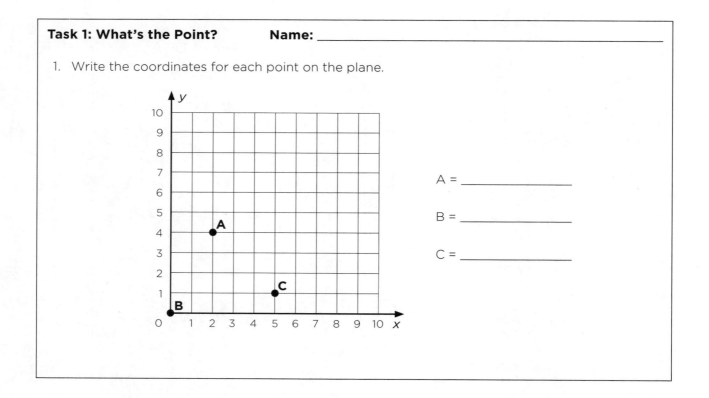

A = _____

B = _____

C = _____

2. Locate the coordinate point that will make a rectangle with points A, B, and C. Name the point D and write the coordinates for D.

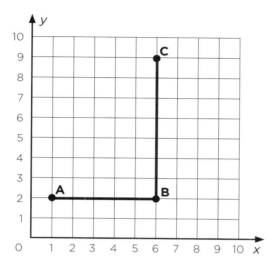

D = _____

3. a. Locate a point that will form a right angle with $\overline{EF}$.

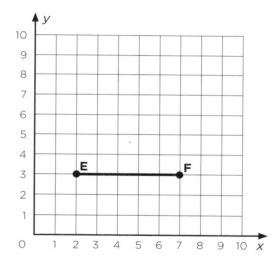

b. Is it possible to find a second point that will create a right angle with $\overline{EF}$? Explain your thinking.

continued →

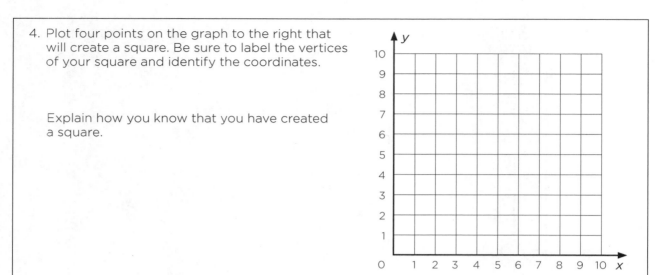

4. Plot four points on the graph to the right that will create a square. Be sure to label the vertices of your square and identify the coordinates.

Explain how you know that you have created a square.

Figure 8.24: Task 1 for Mathematical Practice 8 grade 5 lesson.

*Visit **go.solution-tree.com/MathematicsatWork** for a free reproducible version of this figure.*

Task 2: The Big Cookie Bake   **Name:** _____

Betty Baker loves to bake cookies. She is planning to make lots of cookies for the fifth graders at Julia Child Elementary School. Betty knows that her chunky chocolate chip cookies take eight minutes to bake for each batch and her sugary vanilla cookies take six minutes to bake for each batch.

1. Create a table for each type of cookie that shows the total baking time needed for a given number of batches of cookies.

2. Describe the pattern that you see in each table.

Chunky Chocolate Chip

Number of batches					
Total time (minutes)					

Pattern:

Sugary Vanilla

Number of batches					
Total time (minutes)					

Pattern:

3. Graph the data for both types of cookies on the coordinate plane below. Use a different color to indicate each type of cookie. Be sure to label the graph.

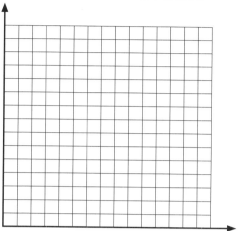

4. After how many minutes will both types of cookies be ready at the same time? How many batches of each cookie will Betty have baked?

5. Are there other times when both types of cookies will be ready to come out of the oven at the same time? How many batches of each cookie type will Betty have baked?

6. Suppose each batch makes twelve cookies. After twenty-four minutes, will Betty have baked enough cookies for every fifth grader at your school to have one of each type of cookie? Explain your thinking.

Figure 8.25: Task 2 for Mathematical Practice 8 grade 5 lesson.

*Visit **go.solution-tree.com/MathematicsatWork** for a free reproducible version of this figure.*

In the beginning-of-class routine, students review plotting points on a coordinate plane. The teacher also reviews attributes of geometric shapes so students can use their understanding as they plot points to create a square when given the location of three of the vertices, as seen in task 1. This opening activity serves to review the use of the coordinate plane and makes a connection to attributes of geometric shapes so students can work together to complete task 1 with limited teacher support.

In task 2, The Big Cookie Bake, students work independently at first to make sense of the question. They begin by building a table for the baking times for each batch of cookies, and they describe the rule that they see in this chart. If you want to make this task less structured, consider removing the chart. You can then see how students establish their own chart, which may provide insights into how they are thinking about the task. Students then graph the data for both kinds of cookies and draw conclusions based on the graph and table. In this task, students are using patterns to solve a problem based on a real-world context.

Summary and Action

By noticing and working with patterns, students are actively involved in building their mathematical understanding. They work to make order out of and memorize random topics. As Arthur Hyde (2006) suggests, "Presenting mathematics as the science of patterns can bring coherence to a bouillabaisse of disconnected ideas" (p. 115).

Standard for Mathematical Practice 8 appears playful as students explore and muck around in the numbers. The work of this practice occurs as students make sense of their findings and explore possible reasons that could make a pattern true. Teachers should consider what generalizations students make about a particular pattern, and extend patterns to future lessons. For example, if a pattern holds true for whole numbers, will it also hold true for fractions?

Identify a content standard you are having students learn currently or in the near future. Choose at least two of the Mathematical Practice 8 strategies from the following list to develop the habits of mind in students in order to look for and express regularity in repeated reasoning.

- Exploring repetition
  - Jump, Stomp, Clap
  - Evens Up
  - Hundreds chart
- Discovering patterns
- Identifying patterns
  - Repeating patterns
  - Growing patterns
- Representing patterns
- Playing games with patterns
  - Poison
  - Tic-tac-toe

Record these in the reproducible "Strategies for Mathematical Practice 8: Look for and Express Regularity in Repeated Reasoning." (Visit **go.solution-tree.com/MathematicsatWork** to download this free reproducible.) How were all students engaged when using the strategy? What was the impact on student learning? How do you know?

EPILOGUE

What Do I Do With These Strategies?

In an excellent mathematics program, educators hold themselves and their colleagues accountable for the mathematical success of every student and for personal and collective professional growth toward effective teaching and learning of mathematics.

— NATIONAL COUNCIL OF TEACHERS OF MATHEMATICS

Throughout this book, you have explored strategies that promote student learning of both the content standards and the Standards for Mathematical Practice. With the emphasis on the Mathematical Practices, students learn to make sense of problems and persevere in solving them (using a variety of strategies purposefully), reason abstractly and quantitatively, and construct and critique viable arguments. Additionally, students learn to look for and use patterns, tools, and models to explain relevant situations, all while being precise with language and notation. With these critical reasoning habits of mind, students become true problem solvers and effective users of mathematics.

Along the way, you have connected best practices in literacy to best practices when teaching mathematics. Continue looking for these links and capitalizing on the successful strategies that any discipline can use to improve student learning and build routines.

Armed with the knowledge that mathematics learning requires much more than rote memorization or reliance on a sequence of steps, the question becomes how to continually grow students' understanding and their ability to apply mathematics. How can students be taught to wonder about this subject and then explore their inquiries? How can students engage in learning during the full mathematics block of time? This book does not give an exhaustive list of strategies to help answer these questions but rather provides a springboard for engagement ideas when teaching content that is related to each Standard for Mathematical Practice.

Instructional Vision

When thinking about the strategies to use in units and lessons, a question quickly arises, What is your vision for instruction? Imagine that your school is the best place for students to learn mathematics and teachers to teach the subject. What would it look like and sound like in the classrooms, hallways, cafeteria, and staff lounge? How would it feel? This becomes your vision, and the instructional decisions you make contribute to that vision.

Use figure EP.1 to make a list of what students and teachers are doing in your vision of quality mathematics instruction and what it looks like, sounds like, and feels like in your school.

	How will it look?	How will it sound?	How will it feel?
Student Actions			
Teacher Actions			

Figure EP.1: Vision for quality mathematics.

*Visit **go.solution-tree.com/MathematicsatWork** for a free reproducible version of this figure.*

Think about how the community views your school. How closely does your vision align to the community perception of mathematics learning at your school? Timothy Kanold (2015c) suggests summing your vision into one sentence of eighteen words or less. What is your vision statement? Suppose you and your colleagues allow yourselves five years to reach your vision. How can you all accomplish your instructional vision within those five years?

The strategies discussed in this book will not, in and of themselves, provide a quick fix to engaging students in learning mathematics or even just learning the content. You must constantly ask yourself: "*Why* am I using this strategy? What will it show me students know and are able to do related to thinking or content understanding? What connections will students make to prior learning when this strategy is employed? How does this strategy reflect the instructional vision?" A strategy for its own sake might be fun or exciting, but if it is not connected to mathematics learning, its purpose is diminished.

To accomplish your vision, quality unit plans and lesson design are critical. How will you use each strategy effectively? When will you be intentionally working with students to learn each habit of mind throughout your units? How will you work collaboratively with your grade-level colleagues to guarantee quality learning for all students in your grade level, regardless of which teacher they have?

Tim Kanold (2015a), as series editor of *Beyond the Common Core: A Handbook for Mathematics in a PLC at Work*, defines a unit as a chunk of content students learn in a two- to four-week time span. Each unit contains content standards students must be proficient with by the end of the unit (though they can always demonstrate proficiency with the standards at a later date) as well as a focus on relevant Mathematical Practices students will simultaneously grow in their understanding and their ability to do.

As you plan for student engagement in your lessons, you and your colleagues must be clear about which standards students are learning and how you will know they are proficient with each one. Although it might be possible to check off or list all eight Mathematical Practices for every unit, focus on those that will be prominent by unit (or lesson). Even when a Standard for Mathematical Practice is checked for a unit, students are never "done" developing that habit of mind. Use figure EP.2 to identify those Mathematical Practices you plan to emphasize and later use to identify potential engaging strategies from the text that match students' expected outcomes for that unit.

	Unit 1	Unit 2	Unit 3	Unit 4	Unit 5	Unit 6	Unit 7	Unit 8	Unit 9	Unit 10
Mathematical Practice 1										
Mathematical Practice 2										
Mathematical Practice 3										
Mathematical Practice 4										
Mathematical Practice 5										
Mathematical Practice 6										
Mathematical Practice 7										
Mathematical Practice 8										

Figure EP.2: Aligning Mathematical Practices to units.

Visit **go.solution-tree.com/MathematicsatWork** *for a free reproducible version of this figure.*

Alternately, your team may want to make a proficiency map similar to the one in figure EP.3 (page 236) to identify those content standards students are expected to be proficient with by the end of each unit and which Mathematical Practices to emphasize while students learn that content.

	Unit 1	Unit 2	Unit 3	Unit 4	Unit 5	Unit 6	Unit 7	Unit 8	Unit 9	Unit 10
Name of Unit										
Number of Days										
Content Standards										
Standards for Mathematical Practice										

Figure EP.3: Proficiency map.

*Visit go.solution-tree.com/**MathematicsatWork** for a free reproducible version of this figure.*

Ideally, this document will be one page so you can see the year at a glance. The name of the unit allows you to see the progression of topics throughout the year to better make connections between the units. The number of days must total less than the total number of instructional days on your district calendar and include assessment and intervention or extension days. Leave room for state assessments, field trips, snow days, and "buffer" days that you might need to add within a unit or between units to re-engage students in learning the priority standards.

Sometimes you'll need to list a content standard in more than one unit. Remember to only list the standard when you expect students to be proficient with it, not when they are solely introduced to it. List it in more than one unit if students will be proficient with a portion of the standard in one unit and the rest of it in another unit. For example, students in first grade might write numbers to 100 in unit 4 and 120 in unit 6.

Plan to implement several Standards for Mathematical Practice at a time, as these help to develop the reasoning and critical-thinking skills students need to successfully understand and apply the content standards.

Armed with a proficiency map or a plan for incorporating student learning of the Standards for Mathematical Practice with the content standards, begin working toward your vision of quality mathematics instruction. Model lessons after those shown in each chapter's lesson-planning tool, and think about the tasks and the strategies that allow all students access to solving those tasks.

Final Reflection

Use figure EP.4 to record the strategies you try for each Mathematical Practice. In the Notes column, identify what you would replicate in future lessons and what you need to modify. If you tried a strategy not on the list, put that in the row titled Other.

Mathematical Practice 1: "Make sense of problems and persevere in solving them."			
Strategy	**Content Learned**	**Notes**	**Date**
Using Graphic Organizers for Building Sense Making			
Reading Word Problems			
Understanding Operations			
Estimating Upfront			
Problem Solving			
Other			

Mathematical Practice 2: "Reason abstractly and quantitatively."			
Strategy	**Content Learned**	**Notes**	**Date**
Quantity Questions			
Graphic Organizers for Reasoning About Quantities			
Always, Sometimes, Never			
Headlines			
Show You Know			
Think-Alouds			
Number Talks			
Other			

Mathematical Practice 3: "Construct viable arguments and critique the reasoning of others."			
Strategy	**Content Learned**	**Notes**	**Date**
Answers First			
What Doesn't Belong?			
Sentence Frames			
Written Explanations			

continued →

Strategy	Content Learned	Notes	Date
Work Comparisons			
Peer Review			
Rubrics			
Student Share			
Other			

Mathematical Practice 4: "Model with mathematics."

Strategy	Content Learned	Notes	Date
Life Observations			
Multiple Representations			
Three Es			
Other			

Mathematical Practice 5: "Use appropriate tools strategically."

Strategy	Content Learned	Notes	Date
Modeling and Incorporating Mathematical Tools			
Toolbox			
Other			

Mathematical Practice 6: "Attend to precision."

Strategy	Content Learned	Notes	Date
Precise Teacher Language			
Precise Student Language			
Symbol Literacy			
Mathematics Vocabulary			

Strategy	Content Learned	Notes	Date
Estimation as a Tool			
Error Analysis			
Calculator Feedback			
Other			

Mathematical Practice 7: "Look for and make use of structure."

Strategy	Content Learned	Notes	Date
Brain Splash			
Algebraic Thinking			
Connections			
Inferences From Structure			
Complex Problems			
Other			

Mathematical Practice 8: "Look for and express regularity in repeated reasoning."

Strategy	Content Learned	Notes	Date
Exploring Repetition			
Discovering Patterns			
Identifying Patterns			
Representing Patterns			
Playing Games With Patterns			
Other			

Figure EP.4: Strategies for the Standards for Mathematical Practice.

*Visit **go.solution-tree.com/MathematicsatWork** for a free reproducible version of this figure.*

Use figure EP.5 (page 240) to record which Mathematical Practices you feel each student has learned and independently demonstrates proficiency with. Note the date you first felt the student demonstrated proficiency. You may want to use this during each quarter or trimester to see how the student develops the Mathematical Practice or as data for your collaborative team to discuss and take action on.

Name	Mathematical Practice 1	Mathematical Practice 2	Mathematical Practice 3	Mathematical Practice 4	Mathematical Practice 5	Mathematical Practice 6	Mathematical Practice 7	Mathematical Practice 8

Consider the following questions.

1. Which Mathematical Practices do you most often emphasize in lessons? How are students demonstrating independence with using those habits of mind?

2. Which Mathematical Practices do you not emphasize often? How can you be sure to provide students with experiences in future units and lessons to build that habit of mind?

Figure EP.5: Proficiency with the Standards for Mathematical Practice.

*Visit **go.solution-tree.com/MathematicsatWork** for a free reproducible version of this figure.*

There are many strategies throughout this resource to engage students in actively exploring and making sense of mathematical ideas. In fact, many of the strong instructional strategies used to teach literacy apply to mathematics as well. Before using any strategy in this book or identifying a high-level task to use with students, always be clear about the content students are learning and then focus on the accompanying Mathematical Practice, strategies, and tasks to move student learning forward. We wish you every success in your journey to engage *all* students as they build their mathematical understanding.

APPENDIX A

Alternate Table of Contents: Literacy Connections

Literacy Connection	Standard for Mathematical Practice	Page Number
Graphic organizers		
• Frayer model	Mathematical Practice 1	20
• Venn diagrams	Mathematical Practice 2	47
• Number webs	Mathematical Practice 2	50
• Rule of Five	Mathematical Practice 4	104
• Vocabulary Organizers	Mathematical Practice 6	162
Key words	Mathematical Practice 1	21
Reading word problems		
• Annotations	Mathematical Practice 1	22
• Highlighting, underlining, and circling	Mathematical Practice 1	24
• Chunking	Mathematical Practice 1	24
• Visualization	Mathematical Practice 1	25
Writing		
• Headlines	Mathematical Practice 2	52
• Three Questions	Mathematical Practice 4	97
• Photographic essay	Mathematical Practice 4	100
• Journal prompt	Mathematical Practice 7	185
Think-alouds	Mathematical Practice 1 Mathematical Practice 2 Mathematical Practice 5	23, 54, 125

Literacy Connection	Standard for Mathematical Practice	Page Number
Stating evidence		
• Always, Sometimes, Never	Mathematical Practice 2	51
• Error Analysis	Mathematical Practice 6	167
Estimating upfront	Mathematical Practice 1	30
Sentence frames	Mathematical Practice 3 Mathematical Practice 8	72, 215
Written explanations	Mathematical Practice 3	74
Peer review	Mathematical Practice 3	80
Rubrics	Mathematical Practice 3	81
Developing vocabulary	Mathematical Practice 6	153
Precise teacher language		
• Teaching academic vocabulary	Mathematical Practice 6	156
• Same Words—Different Meanings	Mathematical Practice 6	156
• Making sense	Mathematical Practice 6	157
• Homophones	Mathematical Practice 6	158
• In measurement	Mathematical Practice 6	158
Precise student language	Mathematical Practice 6	159
Collaborative groups		
• Groups and roles	Mathematical Practice 1	32
• Time to Teach	Mathematical Practice 6	159
Reciprocal teaching		
• Talk-record	Mathematical Practice 1	35
• Write It Right	Mathematical Practice 6	161
Mathematics vocabulary		
• Vocabulary organizers	Mathematical Practice 6	162
• Active word walls	Mathematical Practice 6	163
• Games	Mathematical Practice 6	165
KWL	Mathematical Practice 7	184
Discovering patterns	Mathematical Practice 8	213

APPENDIX B

Alternate Table of Contents: Strategies Building Numeracy and Literacy

If you want to . . .	Standard for Mathematical Practice (Mathematical Practice)	Strategy	Page Number
Increase students' dialogue	Mathematical Practice 1	Shared think-alouds	23
	Mathematical Practice 1	Organizing groups of four students to work on problems and framing their discussions by providing roles	32
	Mathematical Practice 1	Talk-record	35
	Mathematical Practice 3	Sentence frames	72
	Mathematical Practice 3	Student Share	84
	Mathematical Practice 3	What Doesn't Belong?	70
	Mathematical Practice 3	Work comparisons	76
	Mathematical Practice 3	Peer review	80
	Mathematical Practice 6	Reciprocal teaching	159
	Mathematical Practice 6	Back-to-Back	159
Build students' mathematical vocabulary	Mathematical Practice 1	Frayer model	20
	Mathematical Practice 6	Precise teacher language	155
	Mathematical Practice 6	Precise student language	159
	Mathematical Practice 6	Vocabulary organizers	162
	Mathematical Practice 6	Active word walls	163

If you want to . . .	Standard for Mathematical Practice (Mathematical Practice)	Strategy	Page Number
Help students read and interpret problems	Mathematical Practice 1	Key words—dangers	21
	Mathematical Practice 1	Annotations	22
	Mathematical Practice 1	Think-alouds	23
	Mathematical Practice 1	Highlighting	24
	Mathematical Practice 1	Chunking	24
	Mathematical Practice 1	Visualization	25
	Mathematical Practice 1	Estimating upfront	30
	Mathematical Practices 1 & 2	Graphic organizers	17, 47
	Mathematical Practice 2	Headlines	52
	Mathematical Practice 6	Same Words—Different Meanings	156
	Mathematical Practice 7	Break complex problems into smaller problems	194
Understand operations	Mathematical Practice 1	Understanding operations	26
	Mathematical Practice 4	Multiple representations	101
Use estimation	Mathematical Practice 1	Estimating upfront	30
	Mathematical Practice 6	Estimation as a tool	166
Increase problem-solving skills	Mathematical Practice 1	Using graphic organizers for building sense making	17
	Mathematical Practice 1	Problem-solving plan	31
	Mathematical Practice 1	Questions to ask	32
	Mathematical Practice 1	Groups and roles	32
	Mathematical Practice 1	Four Corners	35
	Mathematical Practice 3	Answers First	70
	Mathematical Practice 4	Three Questions	97

If you want to . . .	Standard for Mathematical Practice (Mathematical Practice)	Strategy	Page Number
Ask questions to deepen students' reasoning	Mathematical Practice 2	Always, Sometimes, Never	51
	Mathematical Practice 2	Headlines	52
	Mathematical Practice 2	Number talks	54
	Mathematical Practice 3	Sentence frames	72
	Mathematical Practice 4	Three Es	106
	Mathematical Practice 6	Error analysis	167
	Mathematical Practice 7	Journal prompt	185
	Mathematical Practice 8	Identifying patterns	215
	Mathematical Practice 8	Discovering patterns	213
Provide opportunities to improve students' written expression	Mathematical Practice 2	Headlines	52
	Mathematical Practice 3	Written explanations	74
	Mathematical Practice 3	Answers First	70
	Mathematical Practice 4	Three Questions	97
	Mathematical Practice 4	Photographic essay	100
	Mathematical Practice 6	Precise teacher language	155
	Mathematical Practice 6	Precise student language	159
	Mathematical Practice 6	Write It Right	161
	Mathematical Practice 7	KWL	184
	Mathematical Practice 7	Journal prompt	185
Improve students' listening skills	Mathematical Practice 1	Think-alouds	23
	Mathematical Practice 1	Groups and roles	32
	Mathematical Practice 1	Talk-record	35
	Mathematical Practice 2	Number talks	54
	Mathematical Practice 3	What Doesn't Belong?	70
	Mathematical Practice 3	Work comparisons	76
	Mathematical Practice 6	Reciprocal teaching	159
	Mathematical Practice 6	Back-to-Back	159
	Mathematical Practice 8	Exploring repetition	209

If you want to . . .	Standard for Mathematical Practice (Mathematical Practice)	Strategy	Page Number
Provide opportunities for students to explain their reasoning	Mathematical Practice 2	Quantity questions	46
	Mathematical Practice 2	Always, Sometimes, Never	51
	Mathematical Practice 2	Number talks	54
	Mathematical Practice 3	Answers First	70
	Mathematical Practice 3	What Doesn't Belong?	70
	Mathematical Practice 3	Sentence frames	72
	Mathematical Practice 3	Work comparisons	76
	Mathematical Practice 3	Peer review	80
	Mathematical Practice 3	Find the Error	76
	Mathematical Practice 5	Number lines	132
	Mathematical Practice 8	Discovering patterns	213
Increase students' self-reflection	Mathematical Practice 3	Rubrics	81
	Mathematical Practice 3	Peer review	80
	Mathematical Practice 4	Three Es	106
Make connections with their world	Mathematical Practice 4	Life observations	97
	Mathematical Practice 4	Real-life examples	100
	Mathematical Practice 4	Photographic essay	100
	Mathematical Practice 4	Multiple representations	101
Use multiple representations	Mathematical Practice 1	Studying varied representations using graphic organizers	17
	Mathematical Practice 4	Rule of Five	104
	Mathematical Practice 8	Generalizing patterns with equations, written explanations, or diagrams	215
Use number lines	Mathematical Practice 5	Number lines with counting and rounding	133
	Mathematical Practice 5	Number lines with operations	134
	Mathematical Practice 5	Number lines with fractions	135
	Mathematical Practice 5	Double number lines	135
	Mathematical Practice 5	With measurement and data	135

If you want to . . .	Standard for Mathematical Practice (Mathematical Practice)	Strategy	Page Number
Use technology	Mathematical Practice 5	As a tool	137
	Mathematical Practice 6	Calculator feedback	168
Activate prior knowledge	Mathematical Practice 6	Using graphic organizers	162
	Mathematical Practice 7	Brain Splash	183
	Mathematical Practice 7	KWL	184
	Mathematical Practice 7	Journal prompt	185
	Mathematical Practice 7	Exploration	186
Improve mental strategies	Mathematical Practice 2	Number talks	54
	Mathematical Practice 7	Mental mathematics	188
	Mathematical Practice 7	Connections	189
Use structure	Mathematical Practice 7	Algebraic thinking	188
	Mathematical Practice 7	Connecting solution pathways	190
	Mathematical Practice 7	Inferences from structure	192
	Mathematical Practice 7	Patterns with structure	193
Analyze patterns	Mathematical Practice 7	Inferences from structure	192
	Mathematical Practice 8	Repeating patterns	215
	Mathematical Practice 8	Growing patterns	216
	Mathematical Practice 8	Representing patterns	222
	Mathematical Practice 8	Discovering patterns	213

APPENDIX C

Standards for Mathematical Practice: Background

The Standards for Mathematical Practice were based on two transformative resources, *Principles and Standards for School Mathematics* (NCTM, 2000) and *Adding It Up: Helping Children Learn Mathematics* (Kilpatrick, Swafford, & Findell, 2001). *Principles and Standards for School Mathematics* contributed significantly to the teaching-and-learning-mathematics landscape. NCTM expanded and extended the original 1989 standards to not only include content standards but also process standards. These process standards answered *how* students were to learn, whereas the content standards described *what* K–12 students should learn. The process standards included five key elements.

1. Problem solving

2. Reasoning and proof

3. Communications

4. Connections

5. Representations

These process standards actively engage students in solving problems that enable them to build new knowledge. Through the problem-solving process, students reason, verify their solutions, communicate their ideas, make connections, and use multiple representations of their solutions.

Adding It Up, based on an extensive review of research, presents a definition of mathematical proficiency that includes the following five strands (Kilpatrick et al., 2001).

1. Conceptual Understanding—comprehension of mathematics concepts, operations, and relations

2. Procedural Fluency—skill in carrying out procedures flexibly, accurately, efficiently, and appropriately

3. Strategic Competence—ability to formulate, represent, and solve mathematical problems

4. Adaptive Reasoning—capacity for logical thought, reflection, explanation, and justification

5. Productive Disposition—habitual inclination to see mathematics as sensible, useful, and worthwhile, coupled with a belief in diligence and one's own efficacy. (p. 5)

The process standards and the strands of mathematical proficiency are interdependent and interwoven. The authors of the Common Core State Standards combined these two resources to define the habits of mind that mathematically proficient students demonstrate, and they are evident in the eight Standards for Mathematical Practice (NGA & CCSSO, 2010; see appendix D, page 253).

APPENDIX D

Standards for Mathematical Practice

The Standards for Mathematical Practice describe varieties of expertise that mathematics educators at all levels should seek to develop in their students. These practices rest on important "processes and proficiencies" with longstanding importance in mathematics education. The first of these are the NCTM process standards of problem solving, reasoning and proof, communication, representation, and connections. The second are the strands of mathematical proficiency specified in the National Research Council's report *Adding It Up:* adaptive reasoning, strategic competence, conceptual understanding (comprehension of mathematical concepts, operations and relations), procedural fluency (skill in carrying out procedures flexibly, accurately, efficiently and appropriately), and productive disposition (habitual inclination to see mathematics as sensible, useful, and worthwhile, coupled with a belief in diligence and one's own efficacy).

1. **Make sense of problems and persevere in solving them.** Mathematically proficient students start by explaining to themselves the meaning of a problem and looking for entry points to its solution. They analyze givens, constraints, relationships, and goals. They make conjectures about the form and meaning of the solution and plan a solution pathway rather than simply jumping into a solution attempt. They consider analogous problems, and try special cases and simpler forms of the original problem in order to gain insight into its solution. They monitor and evaluate their progress and change course if necessary. Older students might, depending on the context of the problem, transform algebraic expressions or change the viewing window on their graphing calculator to get the information they need. Mathematically proficient students can explain correspondences between equations, verbal descriptions, tables, and graphs or draw diagrams of important features and relationships, graph data, and search for regularity or trends. Younger students might rely on using concrete objects or pictures to help conceptualize and solve a problem. Mathematically proficient students check their answers to problems using a different method, and they continually ask themselves, "Does this make sense?" They can understand the approaches of others to solving complex problems and identify correspondences between different approaches.

2. **Reason abstractly and quantitatively.** Mathematically proficient students make sense of quantities and their relationships in problem situations. They bring two complementary abilities to bear on problems

involving quantitative relationships: the ability to decontextualize—to abstract a given situation and represent it symbolically and manipulate the representing symbols as if they have a life of their own, without necessarily attending to their referents—and the ability to contextualize, to pause as needed during the manipulation process in order to probe into the referents for the symbols involved. Quantitative reasoning entails habits of creating a coherent representation of the problem at hand; considering the units involved; attending to the meaning of quantities, not just how to compute them; and knowing and flexibly using different properties of operations and objects.

3. Construct viable arguments and critique the reasoning of others. Mathematically proficient students understand and use stated assumptions, definitions, and previously established results in constructing arguments. They make conjectures and build a logical progression of statements to explore the truth of their conjectures. They are able to analyze situations by breaking them into cases, and can recognize and use counterexamples. They justify their conclusions, communicate them to others, and respond to the arguments of others. They reason inductively about data, making plausible arguments that take into account the context from which the data arose. Mathematically proficient students are also able to compare the effectiveness of two plausible arguments, distinguish correct logic or reasoning from that which is flawed, and—if there is a flaw in an argument—explain what it is. Elementary students can construct arguments using concrete referents such as objects, drawings, diagrams, and actions. Such arguments can make sense and be correct, even though they are not generalized or made formal until later grades. Later, students learn to determine domains to which an argument applies. Students at all grades can listen or read the arguments of others, decide whether they make sense, and ask useful questions to clarify or improve the arguments.

4. Model with mathematics. Mathematically proficient students can apply the mathematics they know to solve problems arising in everyday life, society, and the workplace. In early grades, this might be as simple as writing an addition equation to describe a situation. In middle grades, a student might apply proportional reasoning to plan a school event or analyze a problem in the community. By high school, a student might use geometry to solve a design problem or use a function to describe how one quantity of interest depends on another. Mathematically proficient students who can apply what they know are comfortable making assumptions and approximations to simplify a complicated situation, realizing that these may need revision later. They are able to identify important quantities in a practical situation and map their relationships using such tools as diagrams, two-way tables, graphs, flowcharts and formulas. They can analyze those relationships mathematically to draw conclusions. They routinely interpret their mathematical results in the context of the situation and reflect on whether the results make sense, possibly improving the model if it has not served its purpose.

5. Use appropriate tools strategically. Mathematically proficient students consider the available tools when solving a mathematical problem. These tools might include pencil and paper, concrete models, a ruler, a protractor, a calculator, a spreadsheet, a computer algebra system, a statistical package, or dynamic geometry software. Proficient students are sufficiently familiar with tools appropriate for their grade or course to make sound decisions about when each of these tools might be helpful, recognizing both the insight to be gained and their limitations. For example, mathematically proficient high school students analyze graphs of functions and solutions generated using a graphing calculator. They detect possible errors by strategically using estimation and other mathematical knowledge. When making mathematical models, they know that technology can enable them to visualize the results of varying assumptions, explore consequences, and compare predictions with data. Mathematically proficient students at various grade levels are able to identify relevant external mathematical resources, such as digital content located on a website, and use them to pose or solve problems. They are able to use technological tools to explore and deepen their understanding of concepts.

6. Attend to precision. Mathematically proficient students try to communicate precisely to others. They try to use clear definitions in discussion with others and in their own reasoning. They state the meaning of the symbols they choose, including using the equal sign consistently and appropriately. They are careful about specifying units of measure, and labeling axes to clarify the correspondence with quantities in a problem. They calculate accurately and efficiently, express numerical answers with a degree of precision appropriate for the problem context. In the elementary grades, students give carefully formulated explanations to each other. By the time they reach high school they have learned to examine claims and make explicit use of definitions.

7. Look for and make use of structure. Mathematically proficient students look closely to discern a pattern or structure. Young students, for example, might notice that three and seven more is the same amount as seven and three more, or they may sort a collection of shapes according to how many sides the shapes have. Later, students will see 7×8 equals the well remembered $7 \times 5 + 7 \times 3$, in preparation for learning about the distributive property. In the expression $x^2 + 9x + 14$, older students can see the 14 as 2×7 and the 9 as $2 + 7$. They recognize the significance of an existing line in a geometric figure and can use the strategy of drawing an auxiliary line for solving problems. They also can step back for an overview and shift perspective. They can see complicated things, such as some algebraic expressions, as single objects or as being composed of several objects. For example, they can see $5 - 3(x - y)^2$ as 5 minus a positive number times a square and use that to realize that its value cannot be more than 5 for any real numbers x and y.

8. Look for and express regularity in repeated reasoning. Mathematically proficient students notice if calculations are repeated, and look both for general methods and for shortcuts. Upper elementary students might notice when dividing 25 by 11 that they are repeating the same calculations over and over again, and conclude they have a repeating decimal. By paying attention to the calculation of slope as they repeatedly check whether points are on the line through $(1, 2)$ with slope 3, middle school students might abstract the equation $(y - 2)/(x - 1) = 3$. Noticing the regularity in the way terms cancel when expanding $(x - 1)(x + 1)$, $(x - 1)(x^2 + x + 1)$, and $(x - 1)(x^3 + x^2 + x + 1)$ might lead them to the general formula for the sum of a geometric series. As they work to solve a problem, mathematically proficient students maintain oversight of the process, while attending to the details. They continually evaluate the reasonableness of their intermediate results.

Connecting the Standards for Mathematical Practice to the Standards for Mathematical Content

The Standards for Mathematical Practice describe ways in which developing student practitioners of the discipline of mathematics increasingly ought to engage with the subject matter as they grow in mathematical maturity and expertise throughout the elementary, middle and high school years. Designers of curricula, assessments, and professional development should all attend to the need to connect the mathematical practices to mathematical content in mathematics instruction.

The Standards for Mathematical Content are a balanced combination of procedure and understanding. Expectations that begin with the word "understand" are often especially good opportunities to connect the practices to the content. Students who lack understanding of a topic may rely on procedures too heavily. Without a flexible base from which to work, they may be less likely to consider analogous problems, represent problems coherently, justify conclusions, apply the mathematics to practical situations, use technology mindfully to work with the mathematics, explain the mathematics accurately to other students, step back for an overview, or deviate from a known procedure to find a shortcut. In short, a lack of understanding effectively prevents a student from engaging in the mathematical practices.

In this respect, those content standards which set an expectation of understanding are potential "points of intersection" between the Standards for Mathematical Content and the Standards for Mathematical Practice. These points of intersection are intended to be weighted toward central and generative concepts in the school mathematics curriculum that most merit the time, resources, innovative energies, and focus necessary to qualitatively improve the curriculum, instruction, assessment, professional development, and student achievement in mathematics.

APPENDIX E

CCSS for Mathematics Grades K–5

The following table shows the domains and clusters of the content students in grades K–5 must learn in the Common Core State Standards for Mathematics. The Mathematical Practices cannot be taught without content standards. This table shows the content addressed throughout this text when giving strategies and examples for teaching the Mathematical Practices as habits of mind for students to develop.

Table E.1: Grades K–5 Domains and Clusters

	K	1	2	3	4	5
Counting and Cardinality	• Know number names and the count sequence. • Count to tell the number of objects. • Compare numbers.					
Operations and Algebraic Thinking	• Understand addition as putting together and adding to, and understand subtraction as taking apart and taking from.	• Represent and solve problems involving addition and subtraction. • Understand and apply properties of operations and the relationship between addition and subtraction. • Add and subtract within 20. • Work with addition and subtraction equations.	• Represent and solve problems involving addition and subtraction. • Add and subtract within 20. • Work with equal groups of objects to gain foundations for multiplication.	• Represent and solve problems involving multiplication and division. • Understand properties of multiplication and the relationship between multiplication and division. • Multiply and divide within 100. • Solve problems involving the four operations, and identify and explain patterns in arithmetic.	• Use the four operations with whole numbers to solve problems. • Gain familiarity with factors and multiples. • Generate and analyze patterns.	• Write and interpret numerical expressions. • Analyze patterns and relationships.
Number and Operations in Base Ten	• Work with numbers 11–19 to gain foundations for place value.	• Extend the counting sequence. • Understand place value. • Use place value understanding and properties of operations to add and subtract.	• Understand place value. • Use place value understanding and properties of operations to add and subtract.	• Use place value understanding and properties of operations to perform multidigit arithmetic.	• Generalize place value understanding for multidigit whole numbers. • Use place value understanding and properties of operations to perform multidigit arithmetic.	• Understand the place value system. • Perform operations with multidigit whole numbers and with decimals to hundredths.
Number and Operations—Fractions				• Develop understanding of fractions as numbers.	• Extend understanding of fraction equivalence and ordering.	• Use equivalent fractions as a strategy to add and subtract fractions.

	K	1	2	3	4	5
					• Build fractions from unit fractions by applying and extending previous understandings of operations on whole numbers. • Understand decimal notation for fractions, and compare decimal fractions.	• Apply and extend previous understandings of multiplication and division to multiply and divide fractions.
Measurement and Data	• Describe and compare measurable attributes. • Classify objects and count the number of objects in categories.	• Measure lengths indirectly and by iterating length units. • Tell and write time. • Represent and interpret data.	• Measure and estimate lengths in standard units. • Relate addition and subtraction to length. • Work with time and money. • Represent and interpret data.	• Solve problems involving measurement and estimation of intervals of time, liquid volumes, and masses of objects. • Represent and interpret data. • Geometric measurement: understand concepts of area and relate area to multiplication and to addition. • Geometric measurement: recognize perimeter as an attribute of plane figures and distinguish between linear and area measures.	• Solve problems involving measurement and conversion of measurements from a larger unit to a smaller unit. • Represent and interpret data. • Geometric measurement: understand concepts of angle and measure angles.	• Convert like measurement units within a given measurement system. • Represent and interpret data. • Geometric measurement: understand concepts of volume and relate volume to multiplication and to addition.
Geometry	• Identify and describe shapes. • Analyze, compare, create, and compose shapes.	• Reason with shapes and their attributes.	• Reason with shapes and their attributes.	• Reason with shapes and their attributes.	• Draw and identify lines and angles, and classify shapes by properties of their lines and angles.	• Graph points on the coordinate plane to solve real-world and mathematical problems. • Classify two-dimensional figures into categories based on their properties.

Source: NGA & CCSSO, 2010.

APPENDIX F

The Task-Analysis Guide

The following table delineates tasks featured throughout this book as lower- or higher-level-cognitive-demand mathematical tasks.

Table F.1: Cognitive Demand Levels of Mathematical Tasks

Lower-Level Cognitive Demand	Higher-Level Cognitive Demand
Memorization Tasks • These tasks involve reproducing previously learned facts, rules, formulae, or definitions to memory. • They cannot be solved using procedures because a procedure does not exist or because the time frame in which the task is being completed is too short to use the procedure. • They are not ambiguous; such tasks involve exact reproduction of previously seen material and what is to be reproduced is clearly and directly stated. • They have no connection to the concepts or meaning that underlie the facts, rules, formulae, or definitions being learned or reproduced.	**Procedures With Connections Tasks** • These procedures focus students' attention on the use of procedures for the purpose of developing deeper levels of understanding of mathematical concepts and ideas. • They suggest pathways to follow (explicitly or implicitly) that are broad general procedures that have close connections to underlying conceptual ideas as opposed to narrow algorithms that are opaque with respect to underlying concepts. • They usually are represented in multiple ways (for example, visual diagrams, manipulatives, symbols, or problem situations). They require some degree of cognitive effort. Although general procedures may be followed, they cannot be followed mindlessly. Students need to engage with the conceptual ideas that underlie the procedures in order to successfully complete the task and develop understanding.
Procedures Without Connections Tasks • These procedures are algorithmic. Use of the procedure is either specifically called for, or its use is evident based on prior instruction, experience, or placement of the task. • They require limited cognitive demand for successful completion. There is little ambiguity about what needs to be done and how to do it. • They have no connection to the concepts or meaning that underlie the procedure being used. • They are focused on producing correct answers rather than developing mathematical understanding. • They require no explanations or have explanations that focus solely on describing the procedure used.	**Doing Mathematics Tasks** • Doing mathematics tasks requires complex and no algorithmic thinking (for example, the task, instructions, or examples do not explicitly suggest a predictable, well-rehearsed approach or pathway). • It requires students to explore and understand the nature of mathematical concepts, processes, or relationships. • It demands self-monitoring or self-regulation of one's own cognitive processes. • It requires students to access relevant knowledge and experiences and make appropriate use of them in working through the task. • It requires students to analyze the task and actively examine task constraints that may limit possible solution strategies and solutions. • It requires considerable cognitive effort and may involve some level of anxiety for the student due to the unpredictable nature of the required solution process.

Source: Smith & Stein, 1998, as cited in Kanold, 2015a. Reprinted with permission from Mathematics Teaching in the Middle School, *copyright 1998, by the National Council of Teachers of Mathematics. All rights reserved.*

APPENDIX G

Sources for Higher-Level-Cognitive-Demand Tasks

Source: Kanold, 2015a. Used with permission.

Common Core Conversation

www.commoncoreconversation.com/math-resources.html

Common Core Conversation is a collection of more than fifty free website resources for the Common Core State Standards in mathematics and ELA.

EngageNY Mathematics

www.engageny.org/mathematics

The site features curriculum modules from the state of New York that include sample assessment tasks, deep resources, and exemplars for grades preK–12.

Howard County Public School System Secondary Mathematics Common Core

https://secondarymathcommoncore.wikispaces.hcpss.org

This site is a sample wiki for a district K–12 mathematics curriculum.

Illustrative Mathematics

www.illustrativemathematics.org

The main goal of this project is to provide guidance to states, assessment consortia, testing companies, and curriculum developers by illustrating the range and types of mathematical work that students will experience upon implementation of the Common Core State Standards for mathematics.

Inside Mathematics

www.insidemathematics.org/index.php/common-core-standards

This site provides classroom videos and lesson examples to illustrate the Mathematical Practices.

Mathematics Assessment Project

http://map.mathshell.org/materials/index.php

The Mathematics Assessment Project (MAP) aims to bring to life the Common Core State Standards in a way that will help teachers and their students turn their aspirations for achieving them into classroom realities. MAP is a collaboration between the University of California at Berkeley; the Shell Centre team at the University of Nottingham; and the Silicon Valley Mathematics Initiative (MARS).

National Council of Supervisors of Mathematics

www.mathedleadership.org/ccss/itp/index.html

This site features collections of K–12 mathematical tasks for illustrating the Standards for Mathematical Practice. The website includes best-selling books, numerous journal articles, and insights into the teaching and learning of mathematics.

National Council of Teachers of Mathematics Illuminations

http://illuminations.nctm.org

This site provides standards-based resources that improve the teaching and learning of mathematics for all students. The materials illuminate the vision for school mathematics set forth in NCTM's *Principles and Standards for School Mathematics*, *Curriculum Focal Points for Prekindergarten Through Grade 8 Mathematics*, and *Focus in High School Mathematics: Reasoning and Sense Making*.

National Science Digital Library

http://nsdl.org/commcore/math

The National Science Digital Library (NSDL) contains digital learning objects and tasks that are related to specific Common Core State Standards for mathematics.

Partnership for Assessment of Readiness for College and Careers Task Prototypes and New Sample Items for Mathematics

www.parcconline.org/samples/math

This page contains sample web-based practice assessment tasks (released items) for your use.

Smarter Balanced Assessment Consortium Sample Items and Performance Tasks

www.smarterbalanced.org/sample-items-and-performance-tasks

This site contains sample higher-level-cognitive-demand tasks and online test-taking and performance-assessment tasks (released items) for your use in class.

APPENDIX H

Mathematics Vocabulary and Notation Precision

Table H.1 shows common imprecise classroom language with suggestions for precise language (Mathematical Practice 6).

Table H.1: Common Imprecise Language

Imprecise Terms	Precise Terms
Cancel and *canceling*	*Remove common factors* or *divide* if multiplication
Communative or *flip-flop property*	*Commutative* (commute means move)
Plug in	*Substitute*
Using *amount* and *number* interchangeably	*Number* is countable; *amount* is not countable, only measurable.
Using *fewer* and *less* interchangeably	*Fewer* is number; *less* is amount.
Using singulars and plurals interchangeably	*Radius* is singular and *radii* is plural; *die* is singular and *dice* is plural; *parenthesis* is singular and *parentheses* is plural, and *rhombus* is singular while *rhombi* or *rhombuses* are plural.
Pronouncing and spelling singular and plurals for words ending in *x*	*Matrix* and *matrices*, *vertex* and *vertices*, and *index* and *indices*
Using *of* for *times*	*Probability of, logarithm of, function of,* and so on do not mean multiplication.
Points and *corners*	*Vertices*
Timesed or *minused*	*Multiplied* or *subtracted*
Factor out or *pull out*	*Factor to the front—out* indicates that the factor is gone.
Using *and* in whole numbers (2,011 read as "two thousand and eleven")	*And* means a decimal point; 2,011 should be read as "two thousand eleven" as compared to 4.12 as "four *and* twelve hundredths."
Setting up to zero	Set equal to zero
Meters squared, inches squared, feet cubed, and *centimeters cubed*	*Square meters, square inches, cubic feet,* and *cubic centimeters* (units squared or cubed is the notation, not the terminology)
Using *expression* and *equation* interchangeably	*Expression* is used for terms with operations and no equal sign (such as 3 + 5); *equation* means there is an equal sign equating two expressions (such as 4 + ? = 10).

Source: Adapted from Pippen, 2015.

REFERENCES AND RESOURCES

Alexander, R. (2008). *Towards dialogic teaching: Rethinking classroom talk* (4th ed.). York, England: Dialogos.

Baroody, A. (2006). Why children have difficulties mastering the basic number combinations and how to help them. *Teaching Children Mathematics, 13*(1), 22–31.

Boaler, J. (2008). *What's math got to do with it?* New York: Penguin Books.

Boaler, J. (2016). *Mathematical mindsets.* San Francisco: Jossey-Bass.

Burns, M. (1995). *Writing in math class: A resource for grades 2–8.* Sausalito, CA: Math Solutions.

Burns, M. (2007). *About teaching mathematics: A K–8 resource* (3rd ed.). Sausalito, CA: Math Solutions.

Carpenter, T. P., Franke, M. F., & Levi, L. (2003). *Thinking mathematically: Integrating arithmetic and algebra in elementary school.* Portsmouth, NH: Heinemann.

Common Core Standards Writing Team. (2011). *Progressions for the Common Core State Standards in mathematics* [Draft]: *K, counting and cardinality; K–5 operations and algebraic thinking.* Tucson: Institute for Mathematics and Education, University of Arizona. Accessed at https://commoncoretools.files.wordpress.com/2011/05/ccss_progression_cc_oa_k5_2011_05_302.pdf on March 9, 2015.

Common Core Standards Writing Team. (2013). *Progressions for the Common Core State Standards in mathematics* [Draft]. Tucson: Institute for Mathematics and Education, University of Arizona. Accessed at http://ime.math.arizona.edu/progressions on September 17, 2014.

Didax. (2013). *Active place value chart.* Rowley, MA: Author.

Didax. (2015). *Step up to 120 number boards.* Rowley, MA: Author.

Drapeau, P. (2014). *Sparking student creativity: Practical ways to promote innovative thinking and problem solving.* Alexandria, VA: Association for Supervision and Curriculum Development.

Driscoll, M. (1999). *Fostering algebraic thinking: A guide for teachers, grades 6–10.* Portsmouth, NH: Heinemann.

DuFour, R., DuFour, R., Eaker, R., & Many, T. (2010). *Learning by doing: A handbook for Professional Learning Communities at Work* (2nd ed.). Bloomington, IN: Solution Tree Press.

Dweck, C. (2006). *Mindset: The new psychology of success.* New York: Ballantine Books.

EDC Think Math! (n.d.). *Standards for Mathematical Practice: Common Core State Standards for mathematics.* Accessed at www4.uwm.edu/cclm/PDFs/ThinkMath-Math-Practices.pdf on December 31, 2014.

Evertson, C. M., Emmer, E. T., & Worsham, M. E. (2003). *Classroom management for elementary teachers* (6th ed.). Boston: Allyn & Bacon.

Frayer, D. A., Fredrick, W. C., & Klausmeier, H. J. (1969). *A schema for testing the level of concept mastery* (Working Paper No. 16). Madison: Wisconsin Research and Development Center for Cognitive Learning, University of Wisconsin–Madison.

Gillan, S. Y. (1909). *Problems without figures for fourth grade to eighth grade and for mental reviews in high schools and normal schools* (3rd ed.). Milwaukee, WI: Author.

Ginsburg, H. P. (1977). *Children's arithmetic: The learning process.* New York: Van Nostrand.

Hattie, J. (2009). *Visible learning: A synthesis of over 800 meta-analyses relating to achievement.* New York: Routledge.

Hattie, J. (2012). *Visible learning for teachers: Maximizing impact on learning.* New York: Routledge.

Hiebert, J., Carpenter, T. P., Fennema, E., Fuson, K. C., Wearne, D., Murray, H., et al. (1997). *Making sense: Teaching and learning mathematics with understanding.* Portsmouth, NH: Heinemann.

Himmele, P., & Himmele, W. (2011). *Total participation techniques: Making every student an active learner.* Alexandria, VA: Association for Supervision and Curriculum Development.

Hyde, A. (2006). *Comprehending math: Adapting reading strategies to teach mathematics, K–6.* Portsmouth, NH: Heinemann.

Illustrative Mathematics. (2014). *Standards for Mathematical Practice: Commentary and elaborations for K–5.* Tucson, AZ: Author.

Jacobs, H. H. (Ed.). (2010). *Curriculum 21: Essential education for a changing world.* Alexandria, VA: Association for Supervision and Curriculum Development.

Kanold, T. D. (Ed.). (2012a). *Common Core mathematics in a PLC at Work, grades K–2.* Bloomington, IN: Solution Tree Press.

Kanold, T. D. (Ed.). (2012b). *Common Core mathematics in a PLC at Work, grades 3–5.* Bloomington, IN: Solution Tree Press.

Kanold, T. D. (Ed.). (2012c). *Common Core mathematics in a PLC at Work, leader's guide.* Bloomington, IN: Solution Tree Press.

Kanold, T. D. (Ed.). (2015a). *Beyond the Common Core: A handbook for mathematics in a PLC at Work, grades K–5.* Bloomington, IN: Solution Tree Press.

Kanold, T. D. (Ed.). (2015b). *Beyond the Common Core: A handbook for mathematics in a PLC at Work, grades 6–8.* Bloomington, IN: Solution Tree Press.

Kanold, T. D. (Ed.). (2015c). *Beyond the Common Core: A handbook for mathematics in a PLC at Work, leader's guide.* Bloomington, IN: Solution Tree Press.

Kanold, T. D., Briars, D. J., & Fennell, F. (2012). *What principals need to know about teaching and learning mathematics.* Bloomington, IN: Solution Tree Press.

Kelemanik, G. (2015, April 15). *When does 2 + 7 + 8 = 1?* Presentation given at the Math Forum and NCSM Ignite event at the annual conference of the National Council of Supervisors of Mathematics, Boston.

Kenney, P. A., & Kouba, V. L. (1997). What do students know about measurement? In P. A. Kenney & E. A. Silver (Eds.), *Results from the sixth mathematics assessment of the National Assessment of Educational Progress* (pp. 141–164). Reston, VA: National Council of Teachers of Mathematics.

Kilpatrick, J., Swafford, J., & Findell, B. (Eds.). (2001). *Adding it up: Helping children learn mathematics.* Washington, DC: National Academies Press.

Klein, A. S., Beishuizen, M., & Treffers, A. (1998). The empty number line in Dutch second grades: Realistic versus gradual program design. *Journal for Research in Mathematics Education, 29*(4), 443–464.

Lampert, M., & Cobb, P. (2003). Communication and language. In J. Kilpatrick, W. G. Martin, & D. Schifter (Eds.), *A research companion to* Principles and Standards for School Mathematics (pp. 237–249). Reston, VA: National Council of Teachers of Mathematics.

Lappan, G., & Briars, D. J. (1995). How should mathematics be taught? In I. Carl (Ed.), *75 years of progress: Prospects for school mathematics* (pp. 131–156). Reston, VA: National Council of Teachers of Mathematics.

Lucidchart. (2013). *A history of the Venn diagram.* Accessed at www.lucidchart.com/blog/2013/01/17/a-history-of-the-venn-diagram on November 11, 2015.

Marzano, R. J. (2006). *Preliminary report on the 2004–05 evaluation study of the ASCD program for building academic vocabulary.* Alexandria, VA: Association for Supervision and Curriculum Development.

Marzano, R. J., Pickering, D. J., & Pollock, J. E. (2001). *Classroom instruction that works: Research-based strategies for increasing student achievement.* Alexandria, VA: Association for Supervision and Curriculum Development.

McCallum, B. (2011). *Structuring the Mathematical Practices.* Accessed at http://commoncoretools.me/2011/03/10/structuring-the-mathematical-practices on January 4, 2016.

McEwan-Adkins, E. K. (2010). *40 reading intervention strategies for K–6 students: Research-based support for RTI.* Bloomington, IN: Solution Tree Press.

McKeown, M., Beck, I., Omanson, R., & Pople, M. (1985). Some effects of the nature and frequency of vocabulary instruction on the knowledge and use of words. *Reading Research Quarterly, 20*(5), 522–535.

McREL. (2010). *What we know about mathematics teaching and learning* (3rd ed.). Bloomington, IN: Solution Tree Press.

Nagy, W. E., & Scott, J. A. (2000). Vocabulary processes. In M. L. Kamil, P. B. Mosenthal, P. D. Pearson, & R. Barr (Eds.), *Handbook of reading research* (Vol. 3, pp. 269–284). Mahwah, NJ: Erlbaum.

National Council of Supervisors of Mathematics. (2014). *It's TIME: Themes and imperatives for mathematics education.* Bloomington, IN: Solution Tree Press.

National Council of Teachers of Mathematics. (1989). *Curriculum and evaluation standards for school mathematics*. Reston, VA: Author.

National Council of Teachers of Mathematics. (2000). *Principles and standards for school mathematics*. Reston, VA: Author.

National Council of Teachers of Mathematics. (2001). *Navigating through algebra in prekindergarten–grade 2*. Reston, VA: Author.

National Council of Teachers of Mathematics. (2014). *Principles to actions: Ensuring mathematical success for all*. Reston, VA: Author.

National Governors Association Center for Best Practices & Council of Chief State School Officers. (n.d.). *Common Core State Standards for English language arts & literacy in history/social studies, science, and technical subjects: Appendix A—Research supporting key elements of the standards*. Washington, DC: Authors. Accessed at www.corestandards.org/assets/Appendix_A.pdf on September 23, 2015.

National Governors Association Center for Best Practices & Council of Chief State School Officers. (2010). *Common Core State Standards for mathematics*. Washington, DC: Authors. Accessed at www.corestandards.org/assets/CCSSI_Math%20Standards.pdf on September 14, 2014.

National Mathematics Advisory Panel. (2008). *Foundations for success: The final report of the National Mathematics Advisory Panel*. Washington, DC: U.S. Department of Education.

Newaygo County Regional Educational Service Agency. (n.d.). *Marzano's six step process teaching academic vocabulary*. Accessed at www.ncresa.org/docs/PLC_Secondary/_Six_Step_Process.doc on July 15, 2015.

O'Connell, S., & SanGiovanni, J. (2013). *Putting the practices into action: Implementing the Common Core Standards for Mathematical Practice, K–8*. Portsmouth, NH: Heinemann.

Parrish, S. (2010). *Number talks: Helping children build mental math and computation strategies, grades K–5*. Sausalito, CA: Math Solutions.

Perkins, D. (1998). What is understanding? In M. S. Wiske (Ed.), *Teaching for understanding: Linking research with practice* (pp. 39–58). San Francisco: Jossey-Bass. Accessed at www.uwm.edu/~wash/perkins.htm on January 12, 2015.

Pippen, S. (2015). *Imprecise language*. Plainfield, IL: Pippen Consulting.

Pólya, G. (1957). *How to solve it: A new aspect of mathematical method* (2nd ed.). Princeton, NJ: Princeton University Press.

Ray, M. (2013). *Powerful problem solving: Activities for sense making with the mathematical practices*. Portsmouth, NH: Heinemann.

Reinhart, S. C. (2000). Never say anything a kid can say! *Mathematics Teaching in the Middle School, 5*(8), 54–57.

Saphier, J., & Gower, R. (1997). *The skillful teacher: Building your teaching skills* (5th ed.). Acton, MA: Research for Better Teaching.

Schacter, J. (2000). Does individual tutoring produce optimal learning? *American Educational Research Journal, 37*(3), 801–829.

Schuhl, S., & McCaw, S. (2014). *Digging into math.* Newberg, OR: SMc Curriculum.

Schrock, C., Norris, K., Pugalee, D. K., Seitz, R., & Hollingshead, F. (2013). *Great tasks for mathematics K–5.* Denver, CO: National Council of Supervisors of Mathematics.

Seeley, C. L. (2014). *Smarter than we think: More messages about math, teaching, and learning in the 21st century—A resource for teachers, leaders, policy makers, and families.* Sausalito, CA: Math Solutions.

Siena, M. (2009). *From reading to math: How best practices in literacy can make you a better math teacher, grades K–5.* Sausalito, CA: Math Solutions.

Small, M. (2014). *Uncomplicating fractions to meet the Common Core standards in math, K–7.* New York: Teachers College Press.

Smith, M. S. (2000). Redefining success in mathematics teaching and learning. *Mathematics Teaching in the Middle School, 5*(6), 378–382, 386.

Smith, M. S., & Stein, M. K. (1998). Selecting and creating mathematical tasks: From research to practice. *Mathematics Teaching in the Middle School, 3*(5), 344–350.

Smith, M. S., & Stein, M. K. (2011). *5 practices for orchestrating productive mathematics discussions.* Reston, VA: National Council of Teachers of Mathematics.

Smith, M. S., & Stein, M. K. (2012). Selecting and creating mathematical tasks: From research to practice. In G. Lappan, M. K. Smith, & E. Jones (Eds.), *Rich and engaging mathematical tasks, grades 5–9* (pp. 344–350). Reston, VA: National Council of Teachers of Mathematics.

Steen, L. A. (Ed.). (1990). *On the shoulders of giants: New approaches to numeracy.* Washington, DC: National Academies Press.

Stylianides, A. J. (2007). Proof and proving in school mathematics. *Journal for Research in Mathematics Education, 38*(3), 289–321.

Swan, M. (2005). *Standards unit: Improving learning in mathematics—Challenges and strategies.* Nottingham, England: Department for Education and Skills.

Tipps, S., Johnson, A., & Kennedy, L. M. (2011). *Guiding children's learning of mathematics* (12th ed.). Belmont, CA: Wadsworth.

Tyson, K. (2013, May 26). *No tears for tiers: Common Core tiered vocabulary made simple* [Blog post]. Accessed at www.learningunlimitedllc.com/2013/05/tiered-vocabulary on July 15, 2015.

Van de Walle, J. A., & Lovin, L. H. (2006). *Teaching student-centered mathematics, K–3.* Boston: Pearson.

Wiggins, G. (2014, April 23). *Conceptual understanding in mathematics* [Blog post]. Accessed at https://grantwiggins.wordpress.com/2014/04/23/conceptual-understanding-in-mathematics on January 21, 2015.

Wiliam, D. (2011). *Embedded formative assessment.* Bloomington, IN: Solution Tree Press.

INDEX

Beyond the Common Core series
Edited by Timothy D. Kanold
By Thomasenia Lott Adams, Diane J. Briars, Juli K. Dixon, Jessica Kanold-McIntyre, Timothy D. Kanold, Matthew R. Larson, Edward C. Nolan, and Mona Toncheff

Designed to go well beyond the content of your state's standards, this series offers K–12 mathematics instructors and other educators in PLCs an action-oriented guide for focusing curriculum and assessments to positively impact student achievement.
BKF626, BKF627, BKF628, BKF634

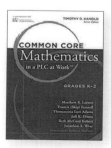

Common Core Mathematics in a PLC at Work®, Grades K–2
Edited by Timothy D. Kanold
By Matthew R. Larson, Francis (Skip) Fennell, Thomasenia Lott Adams, Juli K. Dixon, Beth McCord Kobett, and Jonathan A. Wray

This teacher guide illustrates how to sustain successful implementation of the Common Core State Standards for mathematics, grades K–2. Comprehensive research-affirmed strategies will help your collaborative team develop and assess student demonstrations of deep conceptual understanding and procedural fluency.
BKF566

Common Core Mathematics in a PLC at Work®, Grades 3–5
Edited by Timothy D. Kanold
By Matthew R. Larson, Francis (Skip) Fennell, Thomasenia Lott Adams, Juli K. Dixon, Beth McCord Kobett, and Jonathan A. Wray

This teacher guide illustrates how to sustain successful implementation of the Common Core State Standards for mathematics, grades 3–5. Comprehensive research-affirmed strategies will help your collaborative team develop and assess student demonstrations of deep conceptual understanding and procedural fluency.
BKF568

It's TIME
By the National Council of Supervisors of Mathematics

Help all students become high-achieving mathematics learners. Gain a strong understanding of mathematics culture, and learn necessary best practices to fully align curriculum and instruction with the CCSS for mathematics. You'll explore the factors that have traditionally limited mathematics achievement for students and discover practical strategies for creating an environment that supports mathematics learning and instruction.
BKF600

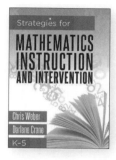

Strategies for Mathematics Instruction and Intervention, K–5
By Chris Weber and Darlene Crane

Build a solid mathematics program by emphasizing prioritized learning goals and integrating RTI into your curriculum. Prepare students to move forward in mathematics learning, and ensure their continued growth in critical thinking and problem solving. With this book, you'll discover an RTI model that provides the mathematics instruction, assessment, and intervention strategies necessary to meet the complex, diverse needs of students.
BKF620

Solution Tree | Press *a division of* Solution Tree

Visit solution-tree.com or call 800.733.6786 to order.

> " I came to the presentation pretty much devoid of an understanding of how the **Common Core** was going to affect my students and my instructional methods. I walked away **excited** and feeling **validated**.
>
> ## I'm on board! "

—David Nohe, teacher,
New Mexico School for the Blind and Visually Impaired

 PD Services

Our experts draw from decades of research and their own experiences to bring you practical strategies for integrating the Common Core. You can choose from a range of customizable services, from a one-day overview to a multiyear process.

Book your CCSS PD today!
888.763.9045

Solution Tree